CONSTRUCTION QUALITY MANAGEMENT

CONSTRUCTION QUALITY MANAGEMENT

S.L. Tang
Syed M. Ahmed
Raymond T. Aoieong
S.W. Poon

HONG KONG UNIVERSITY PRESS

Hong Kong University Press
14/F Hing Wai Centre
7 Tin Wan Praya Road
Aberdeen
Hong Kong

© Hong Kong University Press 2005

ISBN 962 209 746 4

All rights reserved. No part of this publication may be reproduced or transmitted, in any form or by any means, electronic or mechanical, including photocopy, recording, or any information storage or retrieval system, without prior permission in writing from the publisher.

British Library Cataloguing-in-Publication Data
A catalogue record for this book is available from the British Library.

Secure On-line Ordering
http://www.hkupress.org

Printed and bound by Condor Production Co. Ltd., Hong Kong, China

CONTENTS

Preface		vii
Chapter 1:	Introduction to construction quality management	1
Chapter 2:	Quality management and ISO 9000 Standard	15
Chapter 3:	Development of construction quality management in Hong Kong and other countires	29
Chapter 4:	Implementation of quality management systems by Hong Kong contractors and consultants	39
Chapter 5:	Quality indicators	55
Chapter 6:	Quality audits	69
Chapter 7:	Total quality management (TQM) in the construction industry	81
Chapter 8:	Transition from ISO 9000:1994 to ISO 9000:2000 and integration of QMS with TQM philosophy	105
Chapter 9:	Quality cost measurement I (prevention, appraisal and failure costs model)	121

Chapter 10:	Quality cost measurement II (process cost model)	141
Chapter 11:	Applications of CPCM (Construction Process Cost Model)	159
Chapter 12:	Developing a quality culture as the way forward	185
About the Authors		197

PREFACE

The first Quality Assurance Standard, BS 5750, was published in 1979 and the International Organisation for Standardisation used it as the basis for the ISO 9000 standard published in 1987. Over the decade that had passed since BS 5750 (i.e. the 1980s), other industry sectors, particularly the manufacturing sector, had begun to implement these management tools, while there was no apparent interest within the Hong Kong construction industry.

In the 1970s and 1980s, the construction industry in Hong Kong generally adopted the quality control (QC) practice to maintain the standard of construction products. In 1987, the Hong Kong Housing Authority (HKHA) encouraged two major local pre-cast spun concrete pile manufacturers to develop quality management schemes based on the above-said ISO 9000 standard for their products, and hence discovered and began to understand the potential benefits of quality assurance (QA) systems for guaranteeing quality at source. Under the quality surveillance of the manufacturing of concrete piles, it was observed that the quality management schemes provided a high measure of quality assurance for construction processes. Due to the experience gained by HKHA, the Works Bureau of Hong Kong, which undertakes all public civil engineering works in the territory, began to follow HKHA to use the same kind of quality management systems since the 1990s for ensuring the quality of their construction products.

The purpose of this book is to examine the development of quality management systems applied to the construction industry in Hong Kong and other parts of the world. The main emphasis, of course, is based on Hong Kong experience, because at present Hong Kong plays a leading role in construction quality management in the world. Readers will understand how QC (quality control) practice was replaced by QA (quality assurance) concept, and then how the QA concept is being superseded by the TQM (total quality management) philosophy. All the tools and techniques used which are related to construction quality management are discussed in detail throughout the 12 chapters in this book.

Chapter 1 introduces the idea of construction quality management, with definitions relevant terms and an explanation of the differences between QC, QA and TQM. Chapter 2 traces the development of the ISO 9000 standard and highlights the differences between the version in 1994 and that in 2000. In the same chapter, an introduction to quality management systems adopted by construction organizations based on the ISO 9000 standard is also given. Chapter 3 provides a general description of the development and implementation of construction quality management in Hong Kong and other parts of the world, such as Singapore, the UK, the USA, and so on. In Chapter 4, the experience of construction quality management implementation in Hong Kong is discussed in detail. The merits and demerits of implementing such systems experienced by Hong Kong contracting and consulting firms are fully reported in the same chapter.

Chapter 5 introduces what quality indicators, such as CONQUAS and PASS, are. Quality indicators are used to measure the quality of finished products of construction works. Chapter 6 goes on to discuss what quality audits are. Quality audits are used to measure the effectiveness of quality management systems adopted and implemented by construction organizations. Therefore, quality indicators and quality audits serve different purposes, and they are fully discussed in these two chapters. As TQM is a further step of QA implementation, Chapter 7 examines TQM in detail, although it has been generally introduced in Chapter 1. Usually, QA serves as a stepping stone to TQM, the latter being the ultimate goal of quality management. Therefore, after the full account of TQM in Chapter 7, Chapter 8 goes on to describe the Hong Kong construction organizations' experience in transitioning from QA to TQM in their quality management systems, or integrating their quality management systems with the TQM philosophy.

In order to quantify the benefits arise from implementing quality management systems, quality costs must be measurable. In Chapter 9, the definitions of quality costs and the methods of measuring them are given. In the same chapter, the traditional PAF (prevention, appraisal and failure) cost model is described. There are disadvantages in using the PAF model for measuring quality costs of construction works. Chapter 10 proposes an alternative model to the PAF model, called the PCM (process cost model). It has several advantages over the traditional PAF model, and these are discussed in detail in the same chapter. Chapter 11 reports two case studies using the CPCM (construction process cost model) to capture construction quality costs. It can be seen that CPCM is more in line with TQM philosophy and is easy to apply in the construction industry. In Chapter 12, the last chapter, the authors present their personal views on the future development of construction quality management. Some salient ideas, such as quality culture and cultural audits, are thrown out in the chapter. The book ends with a diagram relating several very important aspects of construction quality management: CPCM, quality culture, continual improvement, quality management system and TQM.

As a comprehensive volume covering all the important topics on construction quality management and their latest development, this is a valuable reference book for construction professionals, and is also suitable as a text book for university undergraduate and postgraduate courses in construction management or quality management.

1 INTRODUCTION TO CONSTRUCTION QUALITY MANAGEMENT

Quality is widely accepted as one of the key factors for companies to be successful in the global market. Quality management has been an important issue for many years in various disciplines. The implementation of effective quality management has been witnessed and documented in the manufacturing industry, which set up a paradigm for other disciplines such as the design and construction industry. In the past few years, things have changed in the construction sector. It has opened its doors by welcoming policies that would improve construction process and lead to successful business strategies. Effective quality management, especially **total quality management** (TQM), has been recognized as an enabler for performance improvement in the construction industry.

1.1 An Introduction to Quality

Quality can be defined as a state that meets the legal, aesthetic and functional requirement of a product or project by customers. Requirements may be simple or complex, or they may be stated in terms of the result required or as a detailed description of what is to be done. Metrology, specifications, inspection all go back many centuries before the Christian era. Following the Second World War, two major forces emerged that have had a profound impact on quality: (1) the Japanese revolution in quality; and (2) the prominence of product quality in the public mind. During the twentieth century, a significant body of knowledge on achieving superior quality emerged, advanced by Juran, Deming, Feigenbaum, Crosby and Ishikawa. Here are some examples of defining quality.

"Providing customers with products and services that consistently meet their needs and expectations" — Boeing

"Meeting the customer's need the first time and every time" — General Services Administration, US Government

"Performance to the standard expected by the customer" — FEDEX

"Doing the right thing right the first time, always striving for improvement, and always satisfying the customer" — US Department of Defense

From the above examples, quality has the following common characteristics:
1. It involves meeting or exceeding customer expectations.
2. It applies to products, services, people, processes, and environments.
3. It is an ever-changing state (i.e., what is considered quality today may not be good enough to be considered quality tomorrow).

Quality is a dynamic state associated with products, services, people, processes and environments that meets or exceeds expectations. Quality is always related to customer satisfaction/loyalty and the fitness of use is designed for end customers. Figure 1.1 gives a detailed look into the customer structure of a manufacturing company.

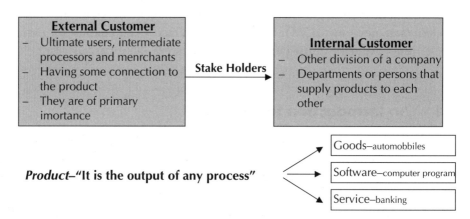

Figure 1.1 Customer structure of a manufacturing company

INTRODUCTION TO CONSTRUCTION QUALITY MANAGEMENT

1.2 The Relationship of Quality with Productivity, Costs, Cycle Time and Value

1.2.1 Quality and Productivity
- Productivity = $\dfrac{\text{Saleable output}}{\text{Resources used}}$
- Improvement in quality directly results in an increase in productivity.

1.2.2 Quality and Costs
- As the quality of design increases, cost increases.
- As the quality of conformance increases, cost decreases.

1.2.3 Quality and Cycle Time
- The cycle time to complete the activities is the key parameter.
- Quality improvement efforts will reduce cycle time.

1.2.4 Quality and Value
- Value = Quality/ Price
- Organizations must evaluate the value they provide, relative to the competition.

1.3 Quality Management

Quality management is the process of identifying and administering the activities needed to achieve the quality objectives of an organization.

| Useful way of introducing qualiy management | ⇒ | Relate it to another well-known management concept (financial) |

Figure 1.2 Diagram of quality management

Quality management is defined as (BS EN ISO 8402):

"All activities of the overall management function that determine the quality policy, objectives and responsibilities, and implement them by means such as quality planning, quality control, quality assurance, and quality improvement within the quality system."

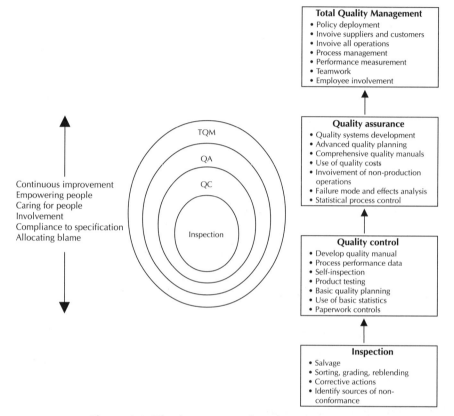

Figure 1.3 The four stages of quality management

The four stages of quality management are shown in the Figure 1.3. Inspection and QC are retrospective; they operate in a detection mode, aiming to find problems that *have occurred* (fire fighting). QA and TQM aim to reduce and ultimately to avoid problems occurring; they can be used to bring about improvement (fire prevention).

1.4 Terminology

Some important terms for quality management are as follows:

Quality

ISO 8402 defines quality as the degree of excellence in a competitive sense, such as reliability, serviceability, maintainability or even individual characteristics.

Quality Control

Both ANSI and ISO define quality control as the operational technique and activity to control and measure the characteristics of a material, structure, component, or system that are used to fulfill requirements for quality.

Quality Systems

Quality systems refer to the organizational structure, procedures, processes and resources needed to implement quality management.

Quality Assurance

Quality assurance is the implementation of planned and systematic activities within quality systems to demonstrate and provide adequate confidence that an entity will fulfill requirements for quality.

Quality Management

Quality management refers to all activities of overall management functions, especially top management leadership, that determine quality policy objectives and responsibilities for all members of the organization.

Total Quality Management

Total quality management is the management approach of an organization, which is based on the participation of its members, concentrates on quality and aims at achieving long-term success through satisfaction and benefits to all members of the organization and society (ISO 8402 and Griffin 1990).

Sometimes, "quality systems" and "quality management" together are described as "quality management systems'. These terms will be used in the rest of this chapter and in the other chapters of this book.

1.5 Quality Evolution

There are five stages in the evolution of quality control, as defined by Rounds and Chi (1984) and Feigenbuam (1991).

- Craftsman quality control was inherent in manufacturing up to the end of the nineteenth century. At that time, a very small number of craftsmen were responsible for the manufacturing of a complete product and each craftsman exclusively controlled the quality of his work.

- Foreman quality control occurred during the industrial revolution when large-scale modern factory concept developed. During this stage, many craftsmen performing similar tasks were grouped together and supervised by a foreman, who then assumed responsibility for the quality of their work.

- Inspection quality control evolved during the First World War when the manufacturing systems became more complex. Because a large number of craftsmen reported to a production foreman, full-time inspectors were required. This era peaked in large organizations in the 1920s and 1930s.

- Statistical quality control flourished during the Second World War when tremendous mass production was necessary. In effect, this step was a refinement of the inspection step and resulted in making the large inspection organizations more efficient. Inspectors were provided with statistical tools such as sampling and control charts. W.A. Shewhart developed a statistical chart for the control of product variables in 1924, marking the beginning of statistical quality control. Later in the same decade, H.F. Hodge and H.G. Roming developed the concept of acceptance sampling as a substitute for 100% inspection; this was considered the most significant contribution of statistical quality control.

- Total quality control evolves in the early 1960s in a four-phase process. A dramatic increase in user quality requirements resulted in increasing customer demand for higher-quality products, forcing the manufacturer to recognize the inadequacy of existing in-plant quality practices and techniques. All these contributed to excessive quality cost, due to such items as inspection, testing, laboratory checks, scrapping and reworking imperfect products, and customer dissatisfaction. These problems highlighted the dual quality challenge: providing significant improvement in the quality of products and practices while, at the same time, effecting substantial reductions in the overall cost of maintaining quality. Statistical quality control could never meet the challenge; thus, a totally new concept was developed based upon the principle that in order to provide genuine effectiveness, control must start with the design of the product and end only when the product has been placed in the hands of a customer who remains satisfied (Feigenbuam 1991).

1.6 Pioneers of the Total Quality Movement

Total quality is not just one individual concept. It is a number of related concepts pulled together to create a comprehensive approach to doing business (Goetsch and Davis 2003). The major contributors to this concept are W. Edwards Deming, Joseph M. Juran, Philip B. Crosby and Kaoru Ishikawa.

1.6.1 Deming

W. Edwards Deming of the United States is considered the father of the quality movement. He is widely known for the Deming Cycle, his Fourteen Points, and the Seven Deadly Diseases.

The Deming Cycle

The basic premise of Deming's cycle covers concepts that apply to all organizations (eCommerce-Now 2001):

1. Plan: How are you going to look at the problem?
2. Do: Carry out the research.
3. Check: Review the results, and see if it achieves what you were aiming to do; if not, go back to step one.
4. Act: Decide that something needs to be altered in your work place, and make a decision to bring about an improvement.

Deming's "plan-do-check-act" cycle will be further discussed in Chapter 7 in detail.

Deming's Fourteen Points

The 14 points are a basis for the transformation of (American) industry. Adoption and action on the 14 points are a signal that management intends to stay in business and aim to protect investors and jobs (Deming 1982, 1986).

1. Create constancy of purpose toward improvement of product and service, with the aim to become competitive and to stay in business, and to provide jobs.
2. Adopt the new philosophy. In a new economic age, management must rise to the challenge, learn their responsibilities, and take on leadership for change.
3. Cease dependence on inspection to achieve quality. Eliminate the need for inspection on a mass basis by building quality into the product in the first place.
4. End the practice of awarding business on the basis of price tag. Instead, minimize total cost. Move towards a single supplier for any one item, on a long term relationship of loyalty and trust.
5. Improve constantly and forever the system of production and service, to improve quality and productivity, and thus constantly decrease costs.
6. Institute training on the job.
7. Institute leadership. The aim of supervision should be to help people and machines and gadgets to do a better job. Supervision of management is in need of an overhaul, as well as supervision of production workers.
8. Drive out fear, so that everyone may work effectively for the company.
9. Break down barriers between departments. People in research, design, sales, and production must work as a team, to foresee problems of production and problems that may be encountered when using the product or service.
10. Eliminate slogans, exhortations, and targets for the workforce asking for zero defects and new levels of productivity. Such exhortations only create adversarial relationships, as the bulk of the causes of low quality and low productivity belong to the system and thus lie beyond the power of the workforce.
11. a. Eliminate work standards (quotas) on the factory floor. Substitute with leadership.
    b. Eliminate management by objective. Eliminate management by numbers, numerical goals. Substitute with leadership.
12. Remove barriers that rob the hourly paid worker of his right to pride in workmanship.
13. Institute a vigorous programme of education and self-improvement.

14. Put everybody in the company to work to accomplish the transformation. The transformation is everybody's job.

Deming's Seven Deadly Diseases
1. Lack of constancy of purpose
2. Emphasis on short-term profits
3. Evaluation by performance, merit rating, or annual review of performance
4. Mobility of management
5. Running a company on visible figures alone
6. Excessive medical costs
7. Excessive costs of warranty, fueled by lawyers that work on contingency fee

1.6.2 Pareto

In 1906, Italian economist Vilfredo Pareto created a mathematical formula to describe the unequal distribution of wealth in his country, observing that 20% of the people owned 80% of the wealth. The 80/20 Rule means that in anything a few (20%) are vital and many (80%) are trivial. In Pareto's case it meant 20% of the people owned 80% of the wealth. In Juran's initial work he identified 20% of the defects causing 80% of the problems. Project managers know that 20% of the work (the first 10% and the last 10%) consumes 80% of the total time and resources. The 80/20 Rule can be applied to almost anything, from the science of management to the physical world.

The value of the Pareto Principle is that it reminds a manager to focus on the 20% that matters. Of all the things done during the day, only 20% really matters. That 20% produces 80% of the total results. Identify and focus on those things. If something in the schedule has to slip, if something is not going to get done, make sure it is not part of that 20% (Reh 2003).

1.6.3 Crosby's Zero Defects

"Zero defects" is a performance standard introduced by Philip H. Crosby in 1979. The Crosby process can help an organization by providing a quality management culture. Lean and Six Sigma can be effective "tools", but a tool becomes more beneficial when it is put to work regularly. To do that, a quality "culture" is needed to encourage (or insist) that everyone participate in this important process (Philip Crosby Ass. 2002).

1.6.4 Ishikawa

The cause and effect diagram is the brainchild of Kaoru Ishikawa, who pioneered quality management processes in the Kawasaki shipyards, and in the process, became one of the founding fathers of modern management. The cause and effect diagram is used to explore all the potential or real causes (or inputs) that result in a single effect (or output). Causes are arranged according to their level of importance or detail, resulting in a depiction of relationships and hierarchy of events. This can help the search for root causes, identify areas where there may be problems, and compare the relative importance of different causes.

1.6.5 Juran

Joseph Juran's "quality trilogy" is made up three components: (1) quality planning, (2) quality control, and (3) quality improvement. This trilogy of quality process leads to successful framework for achieving quality objectives. The processes must occur in an environment of inspirational leadership and the practices must be strongly supportive of quality. A brief description of the Juran's quality trilogy is given below.

1. Quality Planning
   a. Determine who the customers are.
   b. Identify customers' needs.
   c. Develop products with features that respond to customer needs.
   d. Develop systems and processes that allow the organization to produce these features.
   e. Deploy the plans to operational levels.

2. Quality Control
   a. Assess actual quality performance.
   b. Compare performance with goals.
   c. Act on differences between performance and goals.

3. Quality Improvement
   a. Develop the infrastructure necessary to make annual quality improvements.
   b. Identify specific areas in need of improvement, and implement improvement projects.
   c. Establish a project team with responsibility for completing each improvement project.

d. Provide teams with what they need to be able to diagnose problems to determine root causes, develop situations, and establish control that will maintain gains made.

1.7 Quality Control (QC) and Quality Assurance (QA)

Quality control (QC) is the specific implementation of a quality assurance (QA) programme and related activities. Effective QC reduces the possibility of changes, mistakes and omissions, which in turn result in fewer conflicts and disputes.

Quality assurance (QA) is a programme covering activities necessary to provide quality in the work to meet the product/project requirements. QA involves establishing project related policies, procedures, standards, training, guidelines, and system necessary to produce quality. QA provides protection against quality problems through early warnings of trouble ahead. Such early warnings play an important role in the prevention of both internal and external problems.

In practice, the two terms quality assurance and quality control are frequently used interchangeably, which is undesirable. Since quality control is a part of quality assurance, maintaining a clear distinction between then is difficult but important. Quality assurance is all the planned and systematic actions necessary to provide adequate confidence that a structure, system or component will perform satisfactorily and conform to project requirements. On the other hand, quality control is a set of specific procedures involved in the quality assurance process. These procedures include planning, coordinating, developing, checking, reviewing, and scheduling the work. The quality control function is closest to the product in that various techniques and activities are used to monitor the process and to pursue the elimination of sources that lead to unsatisfactory quality performance. Most design-related quality assurance and quality control activities are covered by a design organization's standard office procedures.

1.8 Total Quality Management (TQM)

The primary purpose of TQM is to achieve excellence in customer satisfaction through continuous improvements of products and processes by the total involvement and dedication of each individual who is in any way, a part of that product/process (Ahmed 1993). The principles of TQM create the foundation for developing an organization's system for planning, controlling, and improving quality.

TQM is a structured approach to improvement. If correctly applied, it will assist a construction company in improving its performance. It involves a strong commitment to two guiding principles: customer satisfaction and continuous improvement. In a study of customer satisfaction factors for clients of the transportation, food, chemical and paper, utilities and other miscellaneous industries, it was found that timeliness, cost, quality, client orientation, communication skills, and response to complaints were most significant (Ahmed and Kangari 1995). Another study suggests that TQM methodology like quality function deployment (QFD), provide a structured framework for continuous improvement and customer satisfaction (Ahmed and Kangari 1996).

TQM philosophy will be further discussed in Chapter 7.

References

Ahmed, S. M. (1993). "An Integrated Total Quality Management (TQM) Model for the Construction Process." *Ph.D. Dissertation, School of Civil & Environmental Engineering, Georgia Institute of Technology*, Atlanta, GA, USA.

Ahmed, S. M., and Kangari, R. (1995). "Analysis of Client-Satisfaction Factors in Construction Industry." *Journal of Management in Engineering*, ASCE, 11(2), 36-44

Ahmed, S. M. and Kangari, R. (1996). "Quality Function Deployment in Building Construction." Proceedings of the *2nd International Symposium on QFD*, June 9-11, Novi, Michigan USA, pp. 209-220.

Besterfield D. H. (1994). *Quality Control*. Prentice-Hall.

Beaumont, L.R. (1995). *ISO 9001 The Standard Interpretation*. New Jersey: ISO Easy.

Dale H.B., Carol B., Glen H.B. and Mary B. (2003). *Total Quality Management*, Prentice Hall.

David W. Wood (2003). "TQM makes inroads in construction". http://www.qualitydigest.com/aug/tqm.html

Deming, W. E. (1982). *Quality, Productivity, and competitive position*. Massachusetts Institute of Technology, Cambridge, MA.

Deming, W. E. (1986). *Out of the Crisis*. Massachusetts Institute of Technology, Cambridge, MA.

eCommerce-Now., (2001)., "Deming Quality Cycle," http://www.ecommerce-now.com/images/ecommerce-now/deming.htm

Feigenbaum, Armand V. (1991). *Total Quality Control*. (3rd Edition Revised), McGraw Hill, Singapore.

Goetsch D.L. and Davis S.B. (2003). *Quality Management: Introduction to Total Quality management for Production, Processing, and Services*. 4th Edition, Prentice Hall.

Griffin, Ricky W.(1990) *Management* (3rd edition), Houghton Mifflin Company, Boston.

Harris C. R. (1995). "The Evolution of Quality Management: An overview of the TQM Literature". *Canadian Journal of Administrative Sciences*, 12(2), pp.95-105.

Kanholm, Jack. (2001) ISO 9000:2000 *New Requirements*. Los Angeles: AQA, 2001.

Khosla (2003). "The Deming Management Method".

http://www.khosla.com/softwaremgmt/deming.htm

Kini D. U. (2000). "Global project management—not business as usual." *J. Manage. Eng.*, 16 (6), 29–33.

Krasachol. L. and Tannock J. D. T., (1998), "A Study of TQM Implementation in Thailand". *Proceedings of 3rd International Conference on ISO 9000 and Total Quality Management (Theme: Change for Better)*, School of Business, Hong Kong Baptist University, Hong Kong, April 14-16, 1998, pp. 22-27.

Lew Jeffrey and Hayden Bill Jr., (1992). "Installing TQM Education into the Design and Construction Profession". *Proceedings of ASC 28th Annual Conference*, Auburn University, Auburn, April 9-11, 1992, pp. 35-42.

Love P. E. D., Treloar G. J., Ngowi A. B., Faniran O. O. and Smith J. (2001). "A framework for the Implementation of TQM in Construction Organizations". http://buildnet.csir.co.za/cdcproc/docs/2nd/love_ped.pdf

Lindsey, Patricia and Peoples, Greg (2002). "ISO 9000:2000" *ASCE Proceedings of the 38th Annual Conference*, Virginia Polytechnic Institute and State University, Virginia, April 11 - 13, 2002, pp 293-302

Philip Crosby Associates (2002). "Comparison of Lean Enterprise, Six Sigma and Philip Crosby Quality Management Methodologies".
http://www.philipcrosby.com/pca/C.Articles/articles/year.2002/Comparison.html

Reh, John (2003). "Management Guide" *Legacy USA* http://www.legacyusa.net/pareto.html

Rounds, Gerald L. and Chi, Nai-Yuan (1985), "Total Quality Management for Construction", *ASCE Journal of Construction Engineering and Management*, Vol.111, No.21.

Ho, Samuel K. M. (1998). "Change for the Better via ISO 9000 and TQM," *Proceedings of 3rd International Conference on ISO 9000 and Total Quality Management (Theme: Change for Better)*, School of Business, Hong Kong Baptist University, Hong Kong, April 14-16, 1998, pp. 7-14.

Tenner, Arthur R. and Detoro, Irving J. (1992). *Total Quality Management: Three Steps to Continuous Improvement.* Addison Wiley.

TQM Concept Map (2003).
http://soeweb.syr.edu/faculty/takoszal/TQM.html

Yonatan Reshef (2003). "The Juran Trilogy".
http://courses.bus.ualberta.ca/orga432-reshef/jurantrilolgy.html

2 QUALITY MANAGEMENT AND ISO 9000 STANDARD

2.1 Introduction

Quality has received much attention in construction since the 1990s, or even earlier. Many government departments have made it mandatory for contracting firms to have their quality system accredited. ISO 9000 is the international standard accepted for certification of quality management systems (QMS). While some large contractors are enjoying benefits from implementing their QMS, the smaller firms report difficulties and obstacles.

2.2 Quality Management Systems Adopted by Construction Organizations

According to ISO 9000:2000, a system is a set of interrelated or interacting elements. A system can include different management systems such as a financial management system, an environmental management system and quality management system. For an organization, a quality management system is a management system to direct and control an organization with regard to quality. A construction contractor usually has three quality documents for running a quality management system. The three quality documents are as follows.

2.2.1 Quality Manual

This is a company-wide document setting out the general quality policies, procedures and practices of the organization. A quality manual usually comprises the following:

1. Company policy statement which includes a statement, a summary of activities undertaken and the firm's policy objectives towards implementing a quality system in accordance with the requirements of a standard.
2. General statement to amplify the company's commitment to implementing a quality system.
3. Amendment re-issue and distribution.
4. Authority and responsibility included in the firm's organization.
5. Summary of different procedures.

2.2.2 Quality Procedures

These are documents describing the activities involved in conducting business which are essential to the achievement of quality, e.g. instructions for the production of concrete would require a quality procedure. They are in fact method statements which make reference to relevant specification documents. The quality procedures include the following:

1. Scope and purpose of the procedures.
2. Sequence of actions.
3. Persons responsible in the execution of duties and for ensuring that requirements are met.
4. Remedial actions if non-conformance is detected.

In preparing the quality procedures, the construction firm should already have a number of in-house procedures in controlling its work. Therefore, a substantial part of the preparation of the quality documents entails collecting, documenting and systematizing existing procedures, instructions and practices. The quality documents should be based on the existing practices as long as they are in compliance with the established policies.

2.2.3 Quality Plan

Besides the quality manual and the quality procedures, which are applicable to the entire company, there is also a quality plan which is applicable only to a particular project (or a construction contract) undertaken by the company. Therefore, there can be a number of quality plans for a company, depending on the number of individual projects it is undertaking.

A quality plan is the document derived from the quality system setting out the specific quality practices, resources and activities relevant to a particular contract or project. Normally a quality plan comprises an organization's quality manual, the relevant standard quality procedures and any additional specific quality procedures.

2.3 ISO 9000 Development

The International Organization for Standardization (ISO) was founded in 1946 for the purpose of promoting voluntary, common manufacturing, trade and communication standards. The organization is based in Geneva, Switzerland, and includes 141 member countries and approximately 80 standard-drafting technical committees (Praxiom 2003).

ISO standards are developed according to overall consensus, worldwide industry participation and voluntary efforts of those who apply and operate under a standardized required spectrum. The views of all interests are taken into account. Manufacturers, vendors and users, consumer groups, testing laboratories, governments, engineering professions and research organizations fall within the realm of needing standards. International standardization is market-driven and is not always mandatory by law; therefore it is based on voluntary involvement of all interests in the market place (International Organization for Standardization 2000b).

The International Organization for Standardization (ISO) adopted the ISO 9000 series as the standards for quality management process. It was first released in 1987 and then revised in 1994. These two versions are basically **quality assurance** standards (see Chapter 1). The aim of the ISO 9000 series is to meet a conformance standard, rather than a performance standard. It depends on setting up the formal procedures for all those who are involved to follow. Thus, the work performed is ensured a correct one. The series clarify the distinction and inter-relationships among the principal quality concepts, and provide guidelines on quality systems for both internal and external quality assurance purposes. The standard was further revised in year 2000, which showed some radical changes compared to the 1897 and 1994 versions, because TQM elements can be found in this version.

Although there are several national and international standards developing bodies throughout the world, it is anticipated that ISO 9000 standards will prevail over other standards. ISO 9000 has gained popularity and is being applied to companies and institutions all over the world due to its generic nature. ISO 9000 certification in the construction industry has been widely accepted in many countries, and the number of certifications for general, heavy and specialty contracting companies is growing considerably.

2.4 Quality System Levels (ISO 9000: 1987 version)

ISO 9000 series (1987 version) consists of the following:

ISO 9000

These are the guidelines for selection and use of quality management and quality assurance standards.

ISO 9001

This sets out the model for quality assurance in design, development, production, installation and servicing.

ISO 9002

This sets out the model quality assurance in production, installation and servicing.

ISO 9003

This sets out the model for quality assurance in final inspection and test.

ISO 9004

These are the guidelines for quality management and quality system elements.

Lam *et al.* (1994) lists the reasons of adopting the ISO Series. The Standards are internationally recognized and accepted, hence people involved can speak the same language and compare at the same threshold. The series contain quality assurance criteria which enable quality systems to develop. Further, the Standards provide confidence to customers by third party auditing and certification.

2.5 Quality System Levels (ISO 9000: 1994 Version)

In the 1994 version of ISO 9000 series, three quality system levels are specified which depend on the extent of implementing and application of the quality assurance system.

Level 1 (ISO 9001)
Design, development, production, installation and servicing

ISO 9001 is the most comprehensive quality assurance standard among the three. This level is applied when the technical requirements of a product of service are specified in terms of the performance required. The supplier is responsible for the design, development, manufacture and field trials. As the design is generally not standardized, control of quality throughout all stages is essential to guarantee quality conformance.

Level 2 (ISO 9002)
Production, installation and servicing

This levels applies when the technical requirements can be specified in terms of the standardized or established design, and where quality conformance can be ensured by inspection and test during production.

Level 3 (ISO 9003)
Final inspection and test

This level is applied when quality conformance can be ensured by inspection and tests conducted on the finished product or service.

2.6 Clauses in ISO 9001:1994

The following lists the well-known 20 clauses of ISO 9001 in Section 4 (1994 version) applied to construction works:

4.1 Quality system
All the requirements to establish and maintain a quality system.

4.2 Organization
The delineation of responsibilities as defined in the quality manual.

4.3 Review of the quality system
This ensures effectiveness of the quality system and its suitability to other projects.

4.4 Planning
The logistics of constructing for example the basement of a building.

4.5 Work instructions
The order for activities which affect quality such as mixing and placing of quality.

4.6 Records
A system of records is developed and maintained for inspection and as proof of quality assurance.

4.7 Corrective action
These are procedures to deal with e.g. dimensional discrepancy.

4.8 Design control
A design is produced by following a standard, code of practice or in-house guidelines.

4.9 Documentation and change control
A system to issue and record the amended drawings.

4.10 Control of inspection, measuring and test equipment
The equipment and machines require monitoring for reliability and accuracy.

4.11 Control of purchased material and services
The material and services from suppliers and nominated subcontractor respectively are being nominated.

4.12 Manufacturing control
Control of production on site such as concreting operation.

4.13 Purchaser supplied material
This is to control material and components supplied by the client.

4.14 Completed item inspection and test
Inspection and test of completed work such as the pipeline laid before backfilling.

4.15 Sampling procedures
Procedures applied to control of material such as sampling of concrete for cube tests.

4.16 Control of non-conforming material
This is to deal with non-conforming material such as either rejecting a substandard bored pile or assigning a lower working load to the pile.

4.17 Indication of inspection status
To indicate the inspection results for further action if necessary such as the approved formwork and reinforcement prior to concreting.

4.18 Protection and preservation of production quality
The example is to cover the concrete surface after laying for curing purpose.

4.19 Training
This identifies all training requirements for activities and functions which can affect quality.

4.20 Statistical techniques
This refers to the use of simple charts and diagrams to highlight problems, analyze them and propose various solutions.

2.7   ISO 9000:2000 Version

The 1994 version of the ISO 9000 family has been revised by ISO/TC176 for publication in December 2000. The scope of the standard has been enhanced. ISO 9001:1994 uses the term "quality assurance" and approaches quality assurance by preventing non-conformity. On the contrary, ISO 9001: 2000 uses the term "quality management" and includes meeting customer requirements, regulatory requirements, continual improvement and prevention of non-conformity of a quality management system.

The above mentioned are fundamental changes to the standard which include, inter alia, increased focus on top management commitment, the process approach to quality management, and the move beyond "compliance" towards "customer satisfaction" and "continual improvement" (Lau, 2001). Continual improvement is the process focused on continually increasing the effectiveness and efficiency of the organization to fulfill its policy and objectives. It ensures a dynamic evaluation of the quality management system and responds to the growing needs and expectations of customers.

One objective of ISO 9000:2000 standards is to simplify the structure and reduce the number of standards within the family. This has been achieved by the following:

- The merging of ISO 9001:1994, ISO 9002:1994 and ISO 9003: 1994 by a single quality management system requirements standard, ISO 9001:2000.

- The merging of ISO 8402 and part of the contents of ISO 9000-1 into a new ISO 9000 standard, which is about fundamentals and vocabulary.

- The revision of ISO 9004-1 into a new ISO 9004 standard that provides supplementary guidelines to the new ISO 9001 standard.

The ISO 9000:2000 series is based on eight quality management principles which are derived from the collective experience and knowledge of the internal experts who participated in ISO/TC176. The principles are intended for use by senior management as a framework to guide their organizations towards improved performance. These eight principles include customer focused organization, leadership, involvement of people, process approach, system approach to management, continual improvement, factual approach to decision making and mutually beneficial supplier relationship, which are included in the ISO 9004: 2000 document.

The ISO 9001:2000 version introduced a major change to the structure of the standard, by re-grouping the Clauses 1 to 20 of Section 4 (see 2.6) of the 1994 edition into four sections:

- Section 5 – Management responsibility.
- Section 6 – Resource management.
- Section 7 – Product realization.
- Section 8 – Measurement, analysis and improvement.

The topics of other sections are:

- Section 1 – Scope of the standard.
- Section 2 – Normative reference.
- Section 3 – Definition of terms.
- Section 4 – General and documentation requirements of the quality management system.

It can be seen that the contents of the standard are now organized into a more systematic and readable manner.

The revised ISO 9001:2000 and ISO 9004:2000 standards were developed as a "consistent pair" of standards. The former addresses more clearly the quality management system requirements for an organization to demonstrate its capability to meet customer needs, whereas the latter is intended to lead beyond ISO 9001 towards the development of a comprehensive quality management system, designed to address the needs of all interested parties. Together, the primary aim of the consistent pair is to apply a modern quality management into the processes and activities of an organization, including promotion of continual improvement and achievement of customer satisfaction.

2.8 Developing a Quality Management System

The following are essential stages in developing and implementing a certified quality management system.

Stage 1
The chief executive of the firm should make a commitment to quality assurance by declaring a quality policy such as "Towards total customer satisfaction" and making a formal statement of the objectives such as "To achieve ISO 9001 certification in 15 months' time". Organizing the management structure and defining responsibilities then follow.

Stage 2
Examine and review the existing internal documentation, activities and procedures prior to preparation of quality manual and quality procedures, that is, the quality management system.

Stage 3
When the quality management system is completed and fully approved internally, apply the general quality procedures to specific contracts. Staff should be familiar with quality management and understand their roles. Usually, training would be required.

Stage 4
For specific contracts, the firm has to prepare quality plans and additional quality procedures.

Stage 5
The quality plan is applied to the specific contract and further training may be required.

Stage 6
Internal and external audit (see Chapter 6) of the quality system are being implemented. All quality manuals and quality procedures should be reviewed periodically.

2.9 Objectives, Benefits and Problems of Implementing the QA Systems

Many studies have identified the following objectives, benefits and problems of QA implementation.

Objectives:
- Satisfy the customer's needs and expectations.
- Satisfy the company's needs and interests.

Benefits:
- Reduce wastage.
- Enhance team spirit, instil discipline in staff.
- Reduce staff conflict, enhance job satisfaction.
- Increase efficiency.
- Provide confidence to clients, reduce customer complaints, lower rejection rates, lessen reworks.
- Improve sales.
- Shorten lead time.
- Strengthen relation with subcontractors.
- Lower cost, increase profit.
- Improve systems and standardize procedures.
- Improve workmanship, guarantee quality.

Obstacles and problems faced by the whole construction industry in adopting QA are:
- Short term profit.
- Low priority on quality.
- Tendering at lowest price.
- Multi-layered subcontracting.
- Adversarial relationship in construction industry.
- Lack of training.
- Poor site conditions.
- Complexity and variability of construction process.
- Necessity and ineffectiveness of policing quality.

Problems faced by individual firms:
- Expensive to develop and implement the scheme.
- Lack of right decision by top management.
- Substantial documents and verification.
- Lack of qualified staff.

2.10 Successful and Unsuccessful Case Studies

The implementation of ISO 9000 series would be appropriate to large firms with well defined organization structure and established procedures. A study has revealed that most large construction firms in Hong Kong have their QA schemes accredited while two thirds of the medium sized firms have no such certification. The following reports two cases, a successful one and an unsuccessful one, in setting up the QA systems.

2.10.1 A Successful Case

Contractor A has been established for more than 20 years. This firm has been included in Category Group B of the Government Contractor List and over the years it has completed a lot of small to medium sized building projects.

The main sources of clients are government departments (60%) and one major utility company. This firm is unwilling to take private projects because they have experienced difficulty in getting payment from clients of private projects before.

Under the control of the sole owner, the business has been kept in a stable and smooth condition. The owner has no intention to expand the company rapidly. He always keeps only a few projects in hand fearing that handling too many projects at the same time will require expansion which affects the existing financial and management condition.

A minimum number of staff has been kept in the head office. Five staff members are employed to take care of administration, accounting, tendering and project management. As the owner has just a couple of projects in hand, he can closely supervise the site activities. He is responsible for part of the documentation work and attends the meetings. Basically the decisions are made by him alone. Although he maintains a close relationship with his site staff, he will not keep them when the project is completed in order to reduce overheads.

Almost all internal communications are conducted verbally. The filing system is simple and it records the external correspondence mainly. The owner claims that the system has high efficiency and few mistakes have been made so far.

Contractor A has his own subcontractor list. The selection of subcontractor is always based on past coordination and relationship. However, the lowest quotations are always given priority.

As the Hong Kong Housing Authority requires all new works contractors to be registered with ISO 9000, Contractor A planned to get the ISO 9000 certificate and was successful after 18 months.

However, the owner does not like the system because it is expensive and complicated. Furthermore it has not improved his company significantly so far. He claimed that he has no choice because government projects make up a large proportion of his jobs. Basically he has no expectation on getting more profit in future since normally tenders of the lowest price will be accepted. On the other hand the company will suffer from increasing overheads due to the employment of additional staff to handle the quality documents.

After implementing the system, the owner found that he was not accustomed to the new system. In addition he observed that efficiency was also lowered. Therefore he decided to keep his previous working method but appointed a quality manager to handle the quality document. So, it can be seen that even if an organization is successful in setting up a quality management system and get it certified, it does not mean that the organization must benefit from the system.

2.10.2 An Unsuccessful Case

Contractor B was established in the 1960s. It is a typical family business with members of the family controlling the main positions within the company organization. It handles a wide range of projects including building construction, site formation works, demolition and scaffolding. Investment in China includes running a plant maintenance factory and importing construction material from China to Hong Kong.

The main source of clients comes from the private sector which accounts for 90% of all the projects. The company has developed and maintained good relation with developers, architects and main contractors. About 10 to 20 projects are handled each year with a turnover of about HK$50 million.

The basic strategy of the company is to get the projects which will generate profit. Sometimes it performs as a main contractor, and it is also employed as a subcontractor in large projects. Different trades are sublet and the company will control the site management.

Although the company undertakes a wide range of work, the job description of staff is unclear. They may be required to carry out different types of work. No standardization of duty has been specified and the degree of delegations varies with different projects.

The company has about 25 members of staff and has maintained a family like relationship. Most of the staff will be kept even after site work has been completed. Internal communication and instruction are transmitted verbally. A simple documentation system records the external communications and correspondences.

Purchasing is centralized and controlled by the head office. Selection of the supplier and subcontractor is mainly based on the lowest price, whereas service and payment terms are secondary. The format of the subcontract is simple and sometimes there are verbal contracts.

When the Hong Kong Housing Authority announced that contractors are required to be certified to ISO 9000, the owner planned to get the registration in the hope that it can build up the company's image and get more projects. Furthermore he also wished to improve the company's management system.

A quality consultant was appointed to set up the quality system. Finally a time schedule and cost plan were produced. The initial investment was more than HK$0.25 million and the average monthly cost was about HK$15,000 excluding the employment of a quality manager. Eventually the quality assurance registration proposal was suspended.

The owner had the following reasons for the suspension:

1. The company looks for profit return for survival. As the investment and running cost are high, the owner has no confidence that the system can help gain back the investment.
2. In the private sector, good relationship and lowest price of tender are still the dominant factors to acquire the projects. Without the quality assurance certificate at the moment, the company will not lose jobs. On the other hand, the quality assurance certificate will not guarantee that the company can obtain more jobs from the private sector.
3. To sharpen the competitive edge, the company has to reduce the expenses.
4. The QA system is complicated and there will be implementation problems.

Nevertheless, he agreed that the documentation system would minimize faults and he would selectively implement a part of the system.

References

International Organization for Standardization. (1987). ISO 9000:1987 version.

International Organization for Standardization. (1994). ISO 9000:1994 version.

International Organization for Standardization. (2000a). ISO 9000:2000 version.

International Organization for Standardization. (2000b). [On-Line] Available: http://www.iso.ch.

Lam, S.W., Low, C.M., Teng, W.A. and CIDB Singapore. (1994). *ISO 9000 in Construction*. McGraw-Hill, Singapore.

Lau, Eddy W.T. (2001). "Preparing for Re-certification to ISO 9001". *Hong Kong Engineer, The Journal of the Hong Kong Institution of Engineers*, June 2001, pp24-25.

Praxiom Research Group Limited (2003) "ISO 9000:2000 Introduction". http://praxiom.com/iso-intro.htm

Tang, S.L., Poon, S.W., Ahmed, Syed M. and Wong, K.W. (2003). *Modern Construction Project Management*. 2nd Edition. Hong Kong University Press, Hong Kong. Chapter 7.

3 DEVELOPMENT OF CONSTRUCTION QUALITY MANAGEMENT IN HONG KONG AND OTHER COUNTRIES

3.1 Implementation of ISO 9000 in the Construction Industry

As already introduced in Chapter 2, the ISO 9000 series of standards was first released in 1987. Since then it has become the most popular standard in the world. In a survey conducted by the International Organization for Standardization (ISO, 2003) at the end of the year 2002, it was estimated that at least 561,747 ISO 9000 certificates had been awarded in 159 countries, an increase of 10% over the previous year. In the same survey, it was also found that 12.0% of the certificates had been issued to companies related to the construction sector. This percentage had increased steadily from 8.6% in 1998 to 12.0% in 2002. Among a total of 39 listed industrial sectors, the construction sector recorded the highest number of certificates — the first time since 1998. However, when comparing these figures with those from the total of different manufacturing related sectors, the adoption of ISO 9000 in the construction industry has been slower. One main reason for this slow increase is the lack of client led drive; another reason could be the fact that the construction industry is far more complex than the manufacturing industry. This chapter summarizes recent findings regarding the status of ISO implementation in the construction sector in different regions of the world.

3.2 ISO 9000 and Quality Management Development in the UK Construction Industry

In the UK construction industry, the development of ISO 9000 has been considered a continuation of its pre-successor the BS5750 Quality Systems. Since BS5750 was written primarily to meet the needs of the manufacturing industry, there has been an on-going debate as to the suitability of ISO 9000 to the unique conditions of the construction industry (Shammas-Toma *et al.* 1996). A recent survey was conducted by Moatazed-Keivani *et al.* (1999) to examine the implementation of ISO 9000 in the UK construction industry from the viewpoint of the experiences and perceptions of quality assurance managers. The results of the survey indicated that client demand/request was the main reason for the adoption of QA systems by construction firms, rather than a belief in the value of such systems. Most managers confirmed that this client demand/request was by no means universal or mandatory. Other reasons such as marketing advantage and competitive edge were also reported. One of the major impacts of the ISO certification was the raising of consciousness on quality issues among the top management. In contrast, site employees were not so keen on the implementation of ISO 9000 because of their lack of education and training. While most companies had inevitably experienced increases in paperwork and administrative cost, the majority agreed that the ISO 9000 standards had formed the basis for an efficient quality management system in the construction industry and had also provided the foundation and consciousness for a future move to TQM.

3.3 ISO 9000 and Quality Management Development in the US Construction Industry

A study showed that US companies were about five years behind UK companies in being ISO 9000 registered (Krizan 1999). Dudley, the executive vice president of International Certification Services Ltd., explained that it was because ISO 9000 was largely based on the standard of the UK Ministry of Defense, BS 5750. In addition to that, the US construction industry had, up till then, limited involvement in and hardly any influence on the development and implementation of ISO 9000 (Yates and Aniftos 1996). A survey was conducted by Yates and Aniftos to examine the role international standards, particularly ISO 9000, played in the US construction industry. Out of the 540 questionnaires sent out to different types of companies in the construction industry — such as design, construction, and manufacturing

— 138 questionnaires were returned and analyzed. It was found that approximately 50% of those participating in the study were already registered to ISO 9000, but only nine had construction certification, and eight of these were non-US companies. Of the companies that were not ISO 9000 registered, most indicated "not required in their industry" as their reasons. It was also found that less than 25% of clients requested ISO 9000 registration. The above findings clearly indicated that low registration resulted from a lack of client-led drive even though most companies believed that registration would increase their competitiveness in the world market. Similar surveys in US were also conducted by Chini and Valdez (2003), which consisted of two parts. In the first part survey, out of the 54 questionnaires mailed to construction firms certified for ISO 9000, 36 questionnaires were returned and analyzed. Obtaining access to international markets and satisfying customer requirement were again the main motivations for certification. As a result, it was not surprising to see that the most commonly cited benefits were obtaining an advantage over the competition and using ISO 9000 as a marketing tool. The most common difficulty encountered by respondents was related to documentation and the development of the quality manual. A second part survey of construction firms not certified for ISO 9000 was conducted simultaneously. From this, it was concluded that the most important barrier preventing the certification of the sampled firms was the cost and duration of the certification process. Added to these factors was the widespread belief that ISO 9000 did not apply very well to the construction industry.

3.4 ISO 9000 and Construction Quality Management Development in Hong Kong

Quality management is not a new concept to the Hong Kong construction industry. Before the 1990s, quality of construction works in Hong Kong was largely based on the traditional quality control (QC) methods. The concept of QC was to employ inspection and surveillance methods to control the quality of work. As the industry became aware of the shortcomings of the QC systems, emphasis shifted from the "detection of mistakes" to the "prevention of mistakes" — the so called quality assurance (QA) methodology. In March 1990, the Hong Kong Government launched a "quality awareness campaign" and the Housing Authority took the initiative by striving for an improvement in the quality in the construction of all public housing works. The key events in the Hong Kong Government's drive for quality in construction are summarized in Table 3.1 (Chan et al. 2002; Aoieong and Tang 2004). (For PASS and MASS shown in Table 3.1, please refer to Chapter 5 "Quality Indicators"). Since construction projects were dominated by government ones, contractors had no choice but to follow the timetable prescribed by the Government. In addition, since 1996 the Hong Kong Government Works Branch required all its engineering, architectural associated consultants, List I & List II, and Group C contractors to be certified to ISO 9000. By January 1998, all specialist contractors for land piling – Group II were required to obtain ISO 9000 certification. In the same year, ISO 9001 certification requirement was extended to demolition contractors.

In December 2000, a new edition of ISO 9000 was released, as mentioned in Section 2.7 of Chapter 2. The Hong Kong Government required all construction organizations in Hong Kong to obtain re-certification to this new edition by the middle of December 2003 before they could submit tenders for public jobs. This new edition contains elements for TQM.

Today, clients of the Hong Kong construction industry have succeeded in getting the construction contractors accredited to ISO 9000 standards. Although some benefits were experienced by contractors after the implementation of the quality system (Tam 1996), studies have shown that many problems and difficulties are still encountered by them (Tam and Tong 1996; Ahmed et al. 1998). It is not surprising that most studies arrived at the same conclusion: that client-led drive was the main motivation for certification. The main benefits resulting from certification such as clearer work procedures, better documentation, and competitive advantage were reported by Tam (1996). Kumaraswamy and Dissanayaka (2000) conducted another survey to examine both positive and negative outcomes from ISO 9000 certification as perceived by contractors. The top two negative outcomes were "more paperwork" (100%) and "more time spent on management" (85%); while the top two positive outcomes were "more systematic record keeping" (97%) and "improved internal communication" (91%).

3.5 A Comparative Study of QMS in Hong Kong and the US

A general opinion concerning the implementation of QMS in the US and the Hong Kong construction industries have been obtained through a questionnaire survey and several in-depth interviews carried out in 2002 (Ahmed et al., 2005). Mainly due to the lack of initiative and promotion from both clients and governments, construction companies in the US have

Table 3.1 Calendar of key events

Key Dates	Key Events
1986	A total of 26 public housing blocks that have been in use for some 20 years, housing some 70,000 people, had to be redeveloped because of poor construction quality.
1987	HA encouraged the two major local precast prestressed spun concrete pile manufacturers to develop quality schemes for their products.
1989	HA stepped up measures to improve quality by establishing a 'Quality Assurance Committee', commissioning a consultancy to advise the industry on establishing acceptable quality assurance systems and planning its own lists of approved building contractors.
Early 1990	HA identified the need of a formal set of design and construction procedures for their professionals. HK Industry Department was having a quality drive and encouraged HA to go a step further to require ISO 9000 certification.
Feb 1990	HA began developing the Performance Assessment Scoring System (PASS) to measure contractor performance against defined standards, aiming to provide better tendering opportunities to contractors that score higher in the assessments.
Apr 1990	HA established its own lists of approved building contractors for new works and maintenance works, and required all of them to be certified to ISO 9000 by 31 March 1993.
Jan 1991	HA implements the Performance Assessment Scoring System (PASS)
Early 1991	HA implements the Maintenance Assessment Scoring System (MASS).
Aug 1992	HA required concrete suppliers to be certified to ISO 9000.
Mar 1993	HA required all building contractors to be certified to ISO 9000.
Aug 1993	The Housing Department's Construction Branch was committed to be certified to ISO 9000.
1994	HA implements the 1994 version of Performance Assessment Scoring System (PASS)
Oct 1994	HA required electrical, lift and escalator contractors to be certified to ISO 9000.
Oct 1995	HA required fire services and water pumps contractors to be certified to ISO 9000.
Apr 1996	WB required all engineering, architectural and associated consultants to be certified to ISO 9000.
Oct 1996	WB required all List I & II, Group C contractors to be certified to ISO 9000.
April 1997	HA implements the 1997 version of Performance Assessment Scoring System (PASS)
1997	HA introduced 'Laboratory Assessment Scoring System' for laboratories and the 'Building Services Performance Assessment Scoring System' for building services contractors.
Jan 1998	WB required all specialist contractors for land piling – Group II to obtain ISO 9000 certification.
Jan 1998	ISO 9001 certification requirement was extended to piling contractors.
Mar 1998	HA formed the PASS Control Unit (PCU) to carryout an in-depth review of PASS.
Jul 1998	ISO 9001 certification requirement was extended to demolition contractors.
1999	Preferential Tender Award System was introduced by HA for building contacts.
2000	Implementing the plan "Quality Housing: Partnering for Change".
Jan 2001	Construction Industry Review Committee (CIRC) report recommended 'Partnering' as a solution to the HK construction industry's problems.
Jan 2002	'PASS – January 2002 Edition' was promulgated.
Dec 2003	Re-certification by all construction organizations in Hong Kong to ISO 9000: 2000.

HA: HK Housing Authority, WB: HK Government Works Bureau

failed to see the need to obtain the ISO 9000 certification. In Hong Kong, however, the Government's initiatives have resulted in a high percentage of companies having certified to ISO 9000 standards. A majority of respondents in the US have implemented their own "versions" of QA/QC systems. When comparing these systems with the internationally recognized ISO 9000 management system, it is quite difficult to draw a conclusion as to which system is more effective. However, it is believed that those companies that were ISO 9000 certified would have advantages over the others in the highly competitive international market. In general, construction companies in the US are quite comfortable and content with the present way of doing business. It can be reflected in the high percentage of the US respondents' unawareness of the ISO 9000 standard. On the other hand, though the percentage of ISO 9000 registered companies in Hong Kong is high, it is questionable to conclude whether their motivations to obtain the certification are genuine or not. Among the three key emphases (TQM related) of the Year 2000 edition of the ISO 9000 (i.e. customer satisfaction, process approach and continual improvement), respondents in both the US and Hong Kong considered "customer satisfaction" to be of the utmost importance. Moreover, all respondents ranked "management commitment" as the most important element. Thus it can be concluded that the success of any quality management system depends greatly on the strong commitment of top managements and on how customers are valued.

In order to determine the effectiveness of quality management systems, quantifying quality improvement is essential. Different tools are available for measuring quality improvement of construction processes. Among them, most respondents chose "benchmarking", "statistical process control" and "defect cost analysis". Based on the feedback obtained from interviews, the authors strongly felt that these quality measurements were mainly for monitoring and for recording purposes. One must note that such measurements constitute only the first step towards the never-ending cycles of continual process improvement. The objective is to use the results obtained from such measurements to achieve continual improvement and therefore satisfy the customers' ever changing needs and requirements. The PCM (process cost model) proposed in Chapter 10 should be able to achieve this purpose.

3.6 ISO 9000 and Construction Quality Management Development in Other Countries

Not only were US companies behind UK companies in being ISO 9000 registered, it was also the case in Chile and most of the Latin American countries (Serpell 1999). Research was carried out in Chile by Serpell to examine the problems, limitations and benefits experienced in the process of ISO 9000 implementation. Five construction projects were selected for case studies and the personnel of these projects interviewed. Barriers that contractors, project managers, and owner's staff faced in the quality system implementation processes were observed. It was found that a lack of knowledge of the quality system, non-commitment on the part of top management and the need for support from site personnel were the common barriers confronted by all parties. Once mutual trust between the parties was built and knowledge of the quality system gained through the implementation process, benefits such as improved participation and communication on the site, reduction in any reworking, and better documentation were identified. It was also concluded that quality systems could provide a suitable mechanism to improve the relationship between owners and contractors. This additional benefit would definitely be helpful in alleviating the ever-conflicting relationship between the two parties.

The ISO 9000 Quality Assurance was first introduced to Singapore's construction industry by the Construction Industry Development Board (CIDB) in 1992. In 1995, the Singapore government announced that large construction firms and consultancy firms had to secure ISO 9000 certification by 1999. If they failed to do so, they would either be barred from the public sector and their tendering opportunities would be severely restricted. In 1997, five years after the ISO 9000 quality assurance was introduced, a survey was conducted to obtain general feedback from contractors who were certified to ISO 9000 (Low and Yeo 1997). Thirty-nine survey forms were sent to construction firms and 21 responded to the questionnaire survey. In selecting the reasons for seeking ISO 9000 certification, 90% included the reason "to qualify for tendering for public sector projects", while 33% included the reason "to reduce costs of operations". Similar results were obtained in Hong Kong in the survey conducted by Ahmed and Aoieong (1998). The survey also revealed that benefits such as enhanced communication, improved documentation, improved methods of working, and improved quality of work done were experienced by construction firms following ISO 9000 certification. Among these benefits, 100% of the responding firms agreed that improved documentation was the most important benefit achieved. However, as pointed out by Low and Yeo, having improved quality was indeed a benefit, but having improved quality that did not bring about a reduction in operating costs seemed to be contradictory.

3.7 Concluding Remarks

In summary, ISO 9000 has been, and will continue to be, the most widely accepted quality management system in the construction industry around the world. The steady annual increase in the number of certified construction companies reported in the survey conducted by ISO reinforced this observation (International Organization for Standardization 2003). As suggested by the different surveys discussed in this chapter, most companies do agree that ISO 9000 quality management system is an excellent management tool to improve the overall quality of a company. However, notwithstanding the motivation and/or requirement from clients, the successful use of the tool still largely depends on how that tool is used. In other words, ISO registration alone does not lead to a more efficient quality system. Rather, it is the genuine motive, combined with correct interpretation, formulation and implementation of ISO 9000 that yields the expected results. The 1994 version of ISO 9000 was aimed primarily at quality assurance; however the 2000 version is aimed at customer satisfaction through continual improvement. Thus the criterion to measure system effectiveness now includes customer satisfaction and not merely compliance with requirements. This change in focus will definitely have a tremendous impact on those companies that see the ISO certificates only as a "passport to tender" because additional work will be required to demonstrate compliance to these new requirements.

References

Ahmed, S.M. and Aoieong, Raymond T. (1998). "Analysis of Quality Management Systems in the Hong Kong Construction Industry", *Proceedings of the 1st South African International Conference on Total Quality Management in Construction*, Cape Town, South Africa, pp. 125-134.

Ahmed, Syed. M, Aoieong, Raymond, T, Tang, S.L. and Zhang, X.M. (2005). "A comparison of quality management systems in the construction industries of Hong Kong and USA". International Journal of Quality and Reliability Management, Vol. 22, No. 2, pp.149-161.

Ahmed, S.M., Li, K.T. and Saram, D.D. (1998). "Implementing and Maintaining a Quality Assurance System - A Case Study of a Hong Kong Construction Contractor." *Proceedings of the 3rd International Conference, ISO 9000 & Total Quality Management*, Hong Kong Baptist University, Hong Kong, pp. 450-456.

Aoieong, Raymond T. and Tang, S.L. (2004). "Development of Construction Quality Management and the Related Systems in Hong Kong". Proceedings of the International Symposium on Globalisation and Construction, 17-19 Nov 2004, AIT, Bangkok, Thailand, pp.579-589.

Chan, P.C. Albert, Wong, K.W. Francis, Lam, T.I. and Choi, C.W. (2002). "Quality Development in Hong Kong Public Housing". Proceedings of the First International Conference on Construction in the 21st Century (CITC-I) — Challenges andd Opportunities in Management and Technology, 25-26 April 2002, Miami, Florida, USA, pp.289-294.

Chini, Abdol R. and Valdez, Hector E. (2003). "ISO 9000 and the U.S. Construction Industry", *Journal of Management in Engineering*, ASCE, New York, USA, Vol. 19, No. 2, pp. 69-77.

ISO (2003). *The ISO Survey of ISO 9000 and ISO 14001 Certificates – 12th Cycle*, International Organization for Standardization (ISBN 92-67-10377-6), Geneva, Switzerland.

Krizan, William G. (1999). "ISO Registration Creeps Slowly into Construction", *ENR*, McGraw-Hill Inc., New York, USA, Vol. 243, No. 23, pp. 32-34.

Kumaraswamy, Mohan M. and Dissanayaka, Sunil M. (2000). "ISO 9000 and beyond: from a Hong Kong construction perspective", *Construction Management and Economics*, E & FN Spon, London, UK, Vol. 18, pp. 783-796.

Low, S.P. and Yeo, K.C. (1997). "ISO 9000 quality assurance in Singapore's construction industry: an update", *Structural Survey*, MCB University Press, Bradford, West Yorkshire, UK, Vol. 15, No. 3, pp. 113-117.

Moatazed-Keivani, Ramin, Ghanbari-Parsa, Ali R., and Kagaya, Seiichi (1999). "ISO 9000 standards: perceptions and experiences in the UK construction industry", *Construction Management and Economics*, E & FN Spon, London, UK, Vol. 17, No. 1, pp. 107-119.

Serpell, Alfredo (1999). "Integrating quality systems in construction projects: the Chilean case", *International Journal of Project Management*, Butterworths, Guildford, Surrey, UK, Vol. 17, No. 5, pp. 317-322.

Shammas-Toma, M., Seymour, David E., and Clark, Leslie (1996). "The effectiveness of formal quality management systems in achieving the required cover in reinforced concrete", *Construction Management and Economics*, E & FN Spon, London, UK, Vol. 14, No. 4, pp. 353-364.

Tam, C.M. (1996). "Benefits and Costs of the Implementation of ISO 9000 in the Construction Industry of Hong Kong", *Journal of Real Estate and Construction*, Singapore University Press, Singapore, Vol. 6, No. 1, pp 53-66.

Tam, C.M. and Tong, Thomas K.L. (1996). "A Quality Management System in Hong Kong: a Lesson for the Building Industry Worldwide", *Australian Institute of Building Papers*, Canberra, Australia, Vol. 7, pp. 121-131.

Yates, Janet K. and Aniftos, Stylianos C. (1996). "International Standards: The US Construction Industry's Competitiveness", *Cost Engineering*, American Association of Cost Engineers, Morgantown, WV, USA, Vol. 38, No. 7, pp. 32-37.

4

IMPLEMENTATION OF QUALITY MANAGEMENT SYSTEMS BY HONG KONG CONTRACTORS AND CONSULTANTS

Chapter 3 was a general and brief discussion on the implementation of quality management systems in a number of places in the world. This chapter will concentrate on the situation in Hong Kong alone in greater detail.

The descriptions in this chapter are mainly based on two research findings, Kam and Tang (1998) and Tang and Kam (1999) on the implementation of quality management systems by contractors and consultants respectively in Hong Kong.

Since globally Hong Kong is one of the pioneers in developing construction quality management systems, it is useful to know more about Hong Kong's experience.

4.1   ISO 9000 Certification for Contractors

In June 1992, the first ISO 9002 certificate was issued in Hong Kong to a local building contractor (Tam 1993) by the Hong Kong Quality Assurance Agency (HKQAA), a certifying organization in Hong Kong. Five years later, in June 1997, 350 construction-related firms had received ISO 9000 certificates. In order to learn more about ISO 9000 implementation in the Hong Kong construction industry, a survey was carried out in June/August 1997 for certified contractors.

The purposes of the survey were to find out the contractors' motivation for implementing an ISO 9000 quality management system (QMS), evaluate their experience and difficulties in developing and implementing the QMS, and collect views on their maintenance of the QMS. The survey also examined the benefits of operating ISO 9000 by contractors and asked them about any possible improvement to quality management in the Hong Kong construction industry.

A structured questionnaire was sent to one hundred ISO 9000 certified construction companies that were listed in the Hong Kong Government Works Bureau's approved contractors lists; 35 completed questionnaires were received and the response rate 35% could be considered representative.

4.2 Contractors' Views on Having a QMS to ISO 9000

4.2.1 Motivation for Implementing a QMS

Many reasons and benefits had been suggested as to why companies seek ISO certification (HKQAA 1994; Shaw 1995). These reasons were condensed to seven motivations in the survey. Quality managers of those 35 contractors were asked to indicate their original motivations for having a QMS to ISO 9000. The motivations include:

1. To improve the company's quality image in the construction industry.
2. To improve the company's efficiency and management.
3. To resolve the problems with poor quality arising from construction processes.
4. To reduce the failure costs and liability risks.
5. To fulfill the mandate from the Government.
6. To satisfy the demands from the private owners/clients.
7. To be a stepping stone for implementing TQM (total quality management).

Thirty-three respondents (94%) indicated that the requirement from the Government was the most important reason to seek ISO 9000 certification (item 5). The Hong Kong Housing Department and the Works Bureau required their major contractors to be certified to ISO 9000 as a condition for retention in their approved contractors lists after March 1993 and September 1996 respectively (see Chapter 3). The ISO 9000 certificate becomes a "work permit" and no contractor can afford to ignore it. Twenty contractors (57%) were aware of the advantages of improving the company's image (item 1) and management (item 2) by ISO 9000 certification. Only three companies (9%) recognized that the QMS could help in resolving quality problems (item 3) and reducing the failure costs (item 4). Such a low response indicated that insufficient information about quality-related costs on construction sites led to the difficulty in understanding the long-term savings in construction works. Six companies (17%) pointed out that they had sought certification because of the demands from private developers (item 6), mainly quasi-government organizations such as the Mass Transit Railway Corporation. It was noted that private developers had not yet enforced the ISO 9000 certification requirement for private jobs and the quality of those works could not be measured. There was in general a lack of appreciation of the total quality management (TQM) approach in managing construction projects (item 7). Only two companies (6%) considered that the ISO 9000 certification was a good start for TQM.

4.2.2 Suitability of ISO 9000 to Construction Industry

Since ISO 9000 was first developed for use mainly by the manufacturing industry, there was skepticism about its suitability for use by the construction industry. The quality managers were asked whether they were totally satisfied with ISO 9000 for construction activities. The majority of respondents (74%) agreed that ISO 9000 was an adequate quality system to be applied in their companies. As not all clauses in ISO 9000 were directly related to construction activities, e.g. clause 4.20 — statistical techniques, 23% of the respondents marginally agreed to take ISO 9000 as their quality norm. One quality manager (3%) claimed that construction works heavily relied on professional judgment and disagreed that construction processes were controlled by dogmatic quality requirements. If flexibility is allowed for in the quality procedures, then extensive paper work must be reduced, as he reckoned that it needs excessive paper work in implementing a QMS to ISO 9000.

4.2.3 Difficulties Encountered in Implementing a QMS

Fourteen items of difficulties were assumed in the questionnaire and the opinions from the respondents were sought. These items were:

1. Lack of strong senior management involvement
2. Resistance or bad attitude from staff
3. Poor internal/external communication
4. Not fully understanding requirements of ISO 9000
5. Impractical ISO 9000 requirements on site work

6. Absence of well structured quality system and procedures
7. Too much documentation control and records
8. Change in culture
9. Insufficient quality management training for staff
10. Quality documents in English not easy for site staff and workers to follow
11. Site staff used to working under supervision rather than to a procedural manual
12. QMS not being applied to subcontractors
13. Incompatibility of the contractor's QMS with another QMS operated by the resident engineer appointed by the client
14. Aim at maintaining the ISO 9000 certification as a "work permit" but not genuinely seeking quality improvement

Most of the respondents agreed that the change in culture (item 8) required following the implementation of ISO 9000 brought resistance from the staff (item 2) and affected the performance of the QMS. Site staff are used to working under supervision rather than to quality procedures (item 11) and it require time to change their habit. The required quality documentation control and records have generated extra workload for them (item 7) and this is a difficulty yet to be overcome. Since all responding companies were ISO 9000 certified, they had already met and overcome the problems in receiving management commitment (item 1) and understanding the ISO 9000 requirements (item 4), providing quality management training for staff (item 9), and managing subcontractors (item 12) at the QMS development stage. They did not agreed to the assumed difficulties in items 6, 10 and 13. Remarkably, the quality managers were aware of the unhealthy trend in maintaining the certification as a "work permit". It is hoped that they would seek continual improvement in their QMS.

4.2.4 Maintenance of a QMS

After a QMS has been certified to ISO 9000, many companies only follow the documented procedures to run their quality system without considering the maintenance of the system. Maintaining a QMS is critical to continually satisfying client needs and expectations while protecting the company's interests. Six items for the maintenance and continual improvement of the QMS were proposed to the quality managers for agreement in the questionnaire as follows:

1. Strong motivation from quality management department
2. Regular review of the quality manual and procedures for improvement
3. Retaining quality management staff
4. More quality management information and techniques provided by the Government and certifying organizations
5. Benchmarking with other certified companies
6. Plan to implement TQM system

The quality managers concurred that strong motivation for continual improvement to the system should be initiated from the quality management department (item 1). A regular review of quality documents (item 2) and retention of quality management staff (item 3) were also important factors. As there were only a few quality periodicals published in Hong Kong, the respondents only slightly agreed that they could improve the QMS by obtaining information and techniques from the Government and certifying organizations (item 4). They considered that benchmarking with other ISO 9000 certified contractors (item 5) might share experience in implementing the quality system but no body would like to show trade secret to rivals. A considerable number of contractors planned to implement TQM (item 6) for continual quality improvement at the QMS maintenance stage.

4.2.5 Benefits from Operating a QMS to ISO 9000

Since all the responding companies had achieved ISO 9000 certification and implemented the system for a certain period of time, they were asked to compare the benefits they originally expected to achieve and those they actually achieved as a result of obtaining certification to ISO 9000. The items of benefits listed in the questionnaire were:

1. Enhanced the company's quality image
2. Increased client satisfaction
3. Won more contracts
4. Improved administration system between site and head office
5. Reduced management attention for routine matters and site supervision
6. Trimmed the onerous procedures to improve efficiency
7. Reduced the amount of paper work with a better documentation control
8. Improved construction process and site safety
9. Increased certainty of achieving contract requirements and deadlines

10. Achieved saving through a reduction in failures and reworks
11. Improved personal job satisfaction and morale

A major reason of attaining ISO 9000 certification for many construction companies is to increase their profile in the construction industry. All respondents agreed that the company's image had improved (item 1) and the degree of enhancement was close to their expectation. Improvement in administration system between head and site offices (item 4) were experienced by the contractors. The documented procedures had really improved the company's internal communication.

However, most of the respondents at first believed that the ISO 9000 should greatly increase the overall level of client satisfaction (item 2), but the survey results indicated that the increase in client satisfaction was significantly below expectation. This could only be explained by the fact that contractors had started their QMS for only a few years and had not yet reaped the benefits of implementing the QMS. They still had much to learn about how to implement the QMS effectively. All companies also expected that the certification would have a competitive edge in bidding contracts (item 3), but in fact they found that opportunities almost remained the same, because most other contractors were also certified. They believed that a low tender price was still the main factor in winning contracts. In agreement to a low expectation with ISO 9000 in reducing management attention for routine matters and site supervision (item 5), a negative view was given by the respondents in this regard. The reason perhaps was that the internal audit and system reviews had already demanded a lot of time and attention from the management. The documented procedures could not reduce the level of supervision on workers on an ever changing construction site. There was a relatively high expectation that the trimming of the onerous procedures will improve the efficiency of management (item 6), but the contractors perceived the actual benefit gained from this was lower than expected. Moreover, the contractors clearly understood that there would be an increase in the amount of paperwork in ISO 9000 quality system for document control. Their experience showed that the additional paperwork was even far beyond their anticipation.

Most contractors expected their construction processes and site safety to be better after implementing the QMS (item 8). In practice, they had found no impact in these areas. Once the documented procedures were in place, the contractors had anticipated an improvement in achieving contract requirements and deadlines (item 9). The survey result showed that the respondents were aware of the achievement but the level of improvement was low. Similar result was obtained for item 10 as no significant savings through the reduction in construction failure and rework costs by the QMS were achieved. It was believed that there should be savings but the contractors just did not know the actual amount of savings and therefore gave a neutral response. The respondents also had expected personal job satisfaction and morale to slightly increase (item 11) following the implementation of their QMS but the benefits perceived by them were opposite to their expectations. The staff might have much resistance to paperwork and quality audits at the early stage of implementation. It was likely that once the quality procedures were streamlined, they would enjoy the work more and would make fewer complaints.

The overall benefits which contractors had gained as a result of implementing a QMS to ISO 9000 had not been significant. The level of improvement had not lived up to the respondents' original expectation. Some contractors seemed to have a naïve thinking that simply by writing ISO 9000 into their processes the quality of the products they achieved would be dramatically improved. Many contractors failed to reap the benefits because of the wrong attitude in implementing the QMS mainly from the pressure of the clients to seek ISO 9000 certification. Unless the QMS was adequately planned and maintained, the real benefits of an effective system could not be realized.

4.2.6 Other Observations

In the contractors' head offices, they all had quality management departments and employed quality managers. About two thirds of the contractors in Hong Kong employed external quality consultants to assist in setting up their companies' QMS. Internal staff participation in setting up the systems was necessary because the consultants hired were generally lack of good technical knowledge about construction processes. The annual costs of maintaining the companies' quality management departments ranged from 0.2% to 0.7% of the companies' value of work in hand. The departments were responsible for setting up and reviewing, from time to time, the companies' quality manual and for advising site personnel on the preparation of project quality plans for individual projects. They were also responsible for conducting internal quality audits and arranging training for staff about quality matters. In general, the companies found, after the start of implementing their QMS, that they exercised better document control and observed improvement in the companies' internal management systems, although the overall results obtained from implementing the QMS were below expectations.

All the contractors had quality management teams on all of their sites, though some team leaders worked full time or others part time on quality matters. The ratio of the maintenance costs of the quality teams to the value of works ranged from 0.08% to 0.86%. The ratio tended to decrease with increasing value of works. The site quality management teams were responsible for the preparation and reviewing from time to time the project quality plans, monitoring process control and document control, conducting on-site training on quality matters, and arranging internal site quality audits. Most contractors integrated their internal quality management activities with the site inspection procedures required under the contracts. This was a good practice and should be encouraged.

4.3   ISO 9000 Certification for Consulting Firms

At about the same time when the contractors were surveyed (as described in the previous sections), there were nearly 40 engineering consulting firms in Hong Kong that had already achieved ISO 9000 certification. In order to gauge the opinions of engineering consulting firms as to their experience in implementing QMS to ISO 9000, another survey, similar to that for contractors, was carried out in September/November 1997 for certified consulting firms. Thirty-six structured questionnaires were sent to quality managers of engineering consulting firms that were listed in the Hong Kong Government's Consultants' Services Directory and certified by HKQAA. Nineteen completed questionnaires were returned, of which 14 were from firms with overseas parent companies and 5 from local firms. The response rate of 53% of all ISO 9000 certified consulting firms could be considered high and the results could support definitive conclusions.

4.4   Consultants' Views on Having a QMS to ISO 9000

4.4.1   Motivation for Implementing a QMS

In the contractors' survey, seven motivations were identified. In this consulting firms' survey, two more motivations were added in the questionnaire. The nine motivations were:

1. To improve the company's quality image in the construction industry

2. To improve the company's efficiency and management
3. To improve the internal and external communication
4. To resolve the problems with poor quality arising from poor design work
5. To reduce the liability risks and insurance costs
6. To meet the internal policy requirement from the parent company
7. To fulfill the mandate from the Government
8. To satisfy the demands from the private owners/clients
9. To be a stepping stone for implementing total quality management (TQM)

All 19 respondents indicated that the requirement from the Government (item 7) was the prime reason for seeking ISO 9000 certification. The Works Bureau required its major engineering consultants to be certified to ISO 9000 as a condition for retention in its Consultants' Services Directory after March 1996 (see Chapter 3). The requirement of ISO 9000 certification by the Government was a major driving force and had actually pushed the consultancy services in Hong Kong to a new level. Nine respondents (47%) acknowledged the advantage of promoting the firm's image in the construction industry (item 1). Twelve firms (63%) were aware of the benefit of improving the firm's efficiency and management (item 2). Eight out of the 19 respondents (42%) recognized that the certification would make better internal and external communication (item 3). It was remarkable that a very low proportion of respondents (only two firms, representing 11%) considered quality improvement in their business as an important motivation for seeking certification (item 4). Such a low response rate indicated that the function of the QMS which could rectify the design process to avoid design fault had been overlooked.

Only five firms (26%) accepted that a certified QMS would reduce liability risks and insurance costs (item 5). Six firms (31%) reported that certification was due to the internal policy from a parent company (item 6) and one firm (5%) due to demands from private developer (item 8). A lack of understanding in TQM in design offices was found from the survey results (item 9), as only two firms (11%) took the ISO 9000 certification as a stepping stone to TQM.

4.4.2 Suitability of ISO 9000 to Engineering Consultancy

The respondents were asked to rate how relevant each ISO 9000 quality clause was to their business. They considered that most of the 20 clauses were "considerably relevant" to consultancy services except Clauses 4.11

(Control of Inspection, Measuring and Test Equipment) and 4.15 (Handling, Storage, Packaging, Preservation and Delivery), which were of very little relevance. Furthermore, Clauses 4.9 (Serving) and 4.20 (Statistical Techniques) were classified to be of no relevance to their business. Although this was the survey result, the authors of this book reckoned that these four clauses were not entirely irrelevant. Procedures such as the control of software, delivery of building model, post-contract service and statistical analysis of the causes of non-conformances are relevant to these clauses. Certain aspects such as financial control, human resource management and computer-aided design control were indeed not mentioned in ISO 9000 standard. Special quality procedures for these aspects had been added in some consultants' QMS.

The ISO 9000 was initially developed with the manufacturing industry in mind but was now applied to the design and site supervision processes. The consultants' quality managers were asked whether they were totally satisfied with ISO 9000 for consultancy services; 63% agreed that ISO was an adequate quality system applicable to their business. However, over one-third (37%) had reservations as to the suitability for the standard to their services. As the production work (construction) is undertaken by another party (contractor), the design, development and production processes related to ISO 9000 requirements could not be entirely controlled and performed by the consultants. Difficulties in interpreting the standard, e.g. design validation, had frustrated the staff during establishment and implementation of the QMS. The survey results suggested that although the ISO 9000 was adequately applied to consultancy services, there remained rooms for improvement.

4.4.3 Difficulties Encountered in Implementing a QMS

Fourteen items of common difficulties were identified in the consultants' questionnaire which were slightly different from those for the contractors' survey:

1. Lack of strong senior management involvement
2. Resistance or bad attitude from staff
3. Engineers, who are trained to look for quality, often unconvinced that ISO 9000 is the best way to do so
4. Absence of well structured quality system and procedures
5. Lack of efficient communication even under quality procedures requirements
6. Too much documentation control and records

7. Impractical ISO 9000 requirements on consultancy services
8. Not fully understanding the requirements of ISO 9000 by staff
9. Change in culture
10. Insufficient quality management training for staff
11. No co-operation from the client to meet the procedures under project quality plan
12. QMS not being applied to sub-consultants
13. The clients only require the consultants to have ISO 9000 certification but they do not have such quality management knowledge
14. Aim at maintaining the ISO 9000 certification as a "work permit" but not genuinely seeking for quality improvement

The respondents found that the most difficult task was to make their staff understand and accept the ISO 9000 quality standard (item 8). Engineers always claimed that they were trained to look for quality and the QMS could not help improve their design work (item 3). The excessive documentation procedures (item 6) and the impractical ISO 9000 requirements on consultancy services (item 7) had brought resistance from the professionals. Some of them were afraid that design flexibility would be lost when a manual was used. It was believed that a well structured QMS could allow for flexibility in design process and allay their concerns. The change to a quality culture (item 9), together with insufficient quality training (item 10), had made staff disinclined to place reliance on QMS. Lack of strong support from the management (item 1) and ineffective communication (item 5) in the design team sometimes were problems, though not serious.

There was no strong complaint that the clients just required the consultants to have certification but the clients themselves did not have a similar quality management concept (item 13). The respondents did not agree to items 4, 11 and 12, as their QMS had been certified and were already applied to their sub-consultants and clients to a certain extent. Because of the prerequisite from the Government, the consultants' quality managers were aware of the unhealthy aspects of maintaining the certificate as a "work permit" (item 14). The authors of this book hoped that all consulting firms would seek continual improvement in their QMS. In general, even though difficulties had been encountered, the survey result showed that there was no insurmountable problem, in implementing QMS according to ISO 9000, to engineering consulting firms.

4.4.4 Maintenance of a QMS

Having achieved ISO 9000 certification, many firms only run their business in accordance with the certified QMS, hoping to reap the maximum benefits from it automatically. Continual improvement of the QMS, in fact, was of paramount importance for meeting clients' new requirements and expectations and this would protect the firms' interests. Therefore, how to maintain a QMS was important. Six questions in this respect, similar to those in the contractors' survey, were posed to the consultants' quality managers:

1. Strong motivation from quality management department
2. Regularly review the quality manual and procedures for improvement
3. Retain quality management staff
4. More quality management information and techniques provided by the Government and certifying organizations
5. Benchmarking with other certified companies
6. Plan to implement a TQM system

The consultants' quality managers strongly agreed that retention of quality staff (item 3) was the best way to maintain a QMS. Since engineering consulting firms provided knowledge based services, the staff was really the asset of the firm and affected the overall quality performance. A regular review of quality documents (item 2) and strong motivation for continual improvement to the system initiated from their quality management departments (item 1) were also important factors. A benchmarking exercise might be attempted with ISO 9000 certified consultants sharing experience in implementing the QMS (item 5). However, as the project quality plan would form a part of the technical proposal in a bidding consultancy agreement, nobody would like to reveal its expertise to rival firms. The quality managers marginally agreed that they could improve the QMS by obtaining quality management information and techniques from the Government and certifying organizations (item 4). However, it was noted there were very few quality oriented periodicals published in Hong Kong. In Section 4.4.1, it was mentioned that only a very small number of consulting firms had the motivation to implement a TQM for continual quality improvement at the stage when they planned a QMS. The same was also found at the QMS maintenance stage. There was some doubt whether the quality managers were completely satisfied with the ISO 9000 QMS as they only attempted to meet the minimum quality requirement from the Government and did not invest more for upgrading their QMS to TQM.

4.4.5 Benefits from Operating a QMS to ISO 9000

A very similar (with slight modifications) set of questions were posed to the nineteen consultants as those posed to the thirty five contractors. The set contained eleven questions and they are shown below. The respondents were asked to compare their actual benefit obtained with their original expectation on each of the 11 issues.

1. Enhanced the company's quality image
2. Increased client satisfaction
3. Won more agreements/contracts
4. Improved administration system among different functional departments
5. Reduced management attention required for routine matters and supervision
6. Trimmed the onerous procedures to simpler version and improved efficiency and productivity
7. Reduced the amount of paper work with a better documentation control
8. Improved design process and improved management to the site resident staff
9. Increased certainty of achieving contract requirements and deadlines
10. Achieved saving through a reduction in design failures and reworks
11. Improved personal job satisfaction and morale

A major reason for attaining ISO 9000 certification for many consulting firms was to have better quality image (item 1). All respondents concurred that their firms' image had improved as a result of gaining certification and the degree of enhancement was very close to their expectation. Most of the respondents believed that ISO 9000 certification would increase the level of client satisfaction. The survey results explicitly demonstrated that the increase in client satisfaction met the original high expectation. Improvements in the administrative system operating between different functional departments (item 4) were experienced by the consultants. Although the benefits were slightly below expectation, The QMS had really improved the firms' internal communication and eliminated any possible misunderstanding because the responsibilities and authority attached to each post were clarified.

All ISO 9000 certified consultants expected to have advantages in winning more agreements (item 3). Unfortunately, they found that although opportunities had increased, the certification could not help in the competition as other consultants were also certified. Apart from the technical and quality requirements, they believed that the fee proposal was still the main factor in winning agreements. A negative expectation of a reduction in management attention required for routine matters and supervision (item 5) was given by the respondents, and the survey results showed that the reality was even worse. The reason perhaps was that the system review and quality audits themselves required extra effort from the management. The situation would improve when the QMS became stable after some internal audits had been completed. There was already a low expectation that reducing the onerous procedures to a simpler approach would improve efficiency and productivity (item 6), but the actual benefit gained from this was even lower than expected. Moreover, the consultants clearly understood that there would be an increase in the amount of paperwork in ISO 9000 for document control (item 7). With such a warning or understanding, a well structured QMS was prepared and the survey result showed that the additional paperwork was kept under control and the situation was slightly better than anticipated. As the documents and data could be in the form of any types of media, the use of electronic copies would definitely save on paperwork.

Most consultants expected that their design process and resident site staff management (item 8) would be better after implementing the QMS. However, they had found no significant improvement in these areas in practice. Once the documented procedures were established, it was anticipated that an improvement in achieving contract requirements and deadlines would be achieved (item 9). The survey result, however, showed that both expectation and actual achievement were low and the respondents considered that the QMS could not help in this regard. Ironically, the consultants' quality managers reported that there were no positive savings from the reduction in design failures and rework (item 10) achieved by the QMS. It was believed that the quality managers had no track record with regard to quality costs. These managers had expected the personal job satisfaction and morale (item 11) would slightly decrease following the implementation of their QMS but the disadvantage actually experienced by them was twice as bad as expected. The professionals might have stronger resistance to documented procedures, paperwork and quality audit at the early stage of certification. The authors of this book considered that once the quality procedures were well established and recognized, the professionals would find the quality manual a useful handbook and would not be further troubled by the system.

4.5 Observations from the Surveys

The overall benefits which the consultants in Hong Kong had gained as a result of implementing a QMS to ISO 9000 were not significant. Throughout the surveys on both contractors and consultants, respondents indicated that the level of improvement had not lived up to original expectation. Other research (Brecka 1994) indicated that companies certified for five years or longer would have a higher chance to reap greater benefits from the QMS. It was expected the responding firms would gain more benefits as time goes by.

The contractors and consultants in Hong Kong generally accepted the ISO 9000 as a norm to their QMS and this international standard was generally found to be a suitable quality standard to be applied in the construction industry. (ISO 9000:1994 is a quality assurance standard and not a TQM standard). The most common reason for seeking certification was the mandatory requirement from the Government. The "work permit" had pushed the contractors and consultants to jump on the bandwagon without thorough quality planning for their business, and they faced a lot of problems during the implementation and maintenance of their QMS.

Emphasis had been on conformance to requirements instead of focusing on how to enhance the quality of the final products. In a total quality setting, focus would be on how a project could be done better and how the work could be better coordinated and how changes required could be identified early. A problem observed was that changes and involvement of various parties who needed to be coordinated were viewed as documentation problems. Too much energy and concentration were spent on documentation, whereas in a total quality setting, focus would be on improving the process and thus the product. Another problem observed was that the attitude in case of a defect was to complete a non-conformity form and follow-up. In contrast, in a total quality organization, operatives would be motivated and empowered to study the incident and propose how problems could be eliminated which would result in a rigorous and meticulous effort to improve both processes and products.

Proper quality training for the professionals should be required to improve the manner of managing QMS. Civil engineering students and young/graduate engineers should be trained to develop a quality culture, particularly the total quality (TQM) culture. The syllabuses of civil engineering undergraduate courses should include quality elements and then the graduate engineers could equip themselves with quality knowledge in performing their duties.

References

Brecka, J. (1994). "Study finds that gains with ISO 9000 registration increase over time". *Quality Progress*, May 1994, p.18.

HKQAA. (1994). *An Executive Guide to the Use of the International Standard for Quality Systems*, Hong Kong Quality Assurance Agency.

Kam, C.W. and Tang, S.L. (1998). "ISO 9000 for building and civil engineering contractors". *Transactions of Hong Kong Institution of Engineers*, Vol.5, No. 2, pp.6-10.

Shaw, J. (1995). *ISO 9000 Made Simple.* Management Books 2000 Ltd., UK.

Tam, A. (1993). "Shui On scores a first". *Hong Kong Engineer, Journal of Hong Kong Institution of Engineers*, Vol. 21, No. 2, p.34.

Tang, S.L., Ahmed, Syed M., Lam, W.Y. and De Saram, D.D. (1998). "The Practice of Quality Management Systems in Hong Kong Construction Contractor Organizations", Total Quality Management in Construction: Towards Zero Defect, *Proceedings of the lst South African International Conference on Total Quality Management*, South Africa, November 1998, pp.16-24.

Tang, S.L. and Kam, C.W. (1999). "A survey of ISO 9001 implementation in engineering consultancies in Hong Kong". *International Journal of Quality and Reliability Management*, Vol. 16, No. 6, pp.562-574.

Tang, S.L. and Lau, Andrew W.T. (2000). "An investigation of quality management systems at contractors' head offices and construction sites in Hong Kong". *Proceedings of the International Forum on Project Management, National Project Management Committee of China, Xi'an*, China, October 2000, pp.299-304.

5 QUALITY INDICATORS

5.1 Introduction

The ISO 9000 series of standards was released in 1987. In the late 1980s, when some contractors were developing or implementing their quality management systems, the housing and construction authorities in Singapore and Hong Kong initiated systems to assess the quality of constructed works. While the ISO 9000 series provides a model for quality management systems, **quality indicators** are used to measure or assess the quality of finished products. (Quality audits, which will be discussed in Chapter 6, however, measure or assess the degree of success of the implementation of the quality management systems.) This chapter highlights the development and implementation of two such quality indicators.

5.2 Background

During the boom in the Singapore construction industry in the early 1980s, construction was a highly labour-intensive industry. The shortage of skilled workers often resulted in poor quality construction (Lam *et al.* 1994). In 1989, the Construction Industry Development Board (CIDB) set up a measurement system — Construction Quality Assessment System (CONQUAS) — to assess construction quality. At about the same time in

Hong Kong, housing was in great demand and quality problems of building works, particularly the housing construction, also appeared. In 1986, 26 housing blocks were required to be demolished prematurely as a result of very poor concrete quality. Therefore, in 1990, the Hong Kong Housing Authority implemented a quality measurement scheme — the Performance Assessment Scoring System (PASS) — for monitoring the performance of building contractors on new works. PASS was developed based on the CONQUAS of Singapore.

5.3 CONQUAS in Singapore

CONQUAS, first introduced in 1989, was developed by CIDB in conjunction with other major public sector agencies including the Housing & Development Board (HDB), Public Works Department and Port of Singapore Authority. It is a standardized, quantifiable and systematic assessment system to grade the construction quality of completed buildings (finished works). It sets out the standard and criteria to measure the quality of different parts of building work and award scores to the work accordingly.

CONQUAS (1989) assessment consists of three parts:

- Architectural (50%)
- Structural (40%)
- External works (10%)

These ratios roughly represent the cost of the works of a reinforced concrete building. The assessment excludes other works such as foundation, sub-structural works, and electrical and mechanical installation works, which are generally carried out under separate contracts.

Each part of the assessment is further divided into sub-items for precise assessment which can represent the performance of the whole building. The assessment should be carried out by the project engineer, the architect or a professional from an independent party such as CIDB.

The summation of the quality score for architectural, structural and external works will yield the total CONQUAS score for a completed building. This score may be reduced by the defects due to poor workmanship or low quality of materials discovered during the maintenance stage.

The average scores of the whole industry or with respect to a particular type of building construction can be compared on a yearly basis. Thus, whether there is improvement or not in the quality of the building constructed can be identified. On the other hand, the industry can specify a target for the industry to achieve.

CONQUAS was first made compulsory for building projects in the public sector. In 1991 it was applied to the superstructure works of private building projects to be assessed by CIDB with a service charge. From 1993, the assessment has been extended to development on sites sold by the HDB and the Urban Redevelopment Authority as a way to assure quality even in the private sector.

CONQUAS has also been extended to civil engineering works construction. In 1993, the Civil Engineering Quality Assessment System (CE CONQUAS), based on the principles of CONQUAS, was adopted. The assessment covers the construction works in the following five main categories:

- Road works and car parks
- Bridges and flyovers
- Drainage works
- Sewerage works
- Marine structures

Each category of works is broken down into various components for assessment according to quality standards and tolerances as specified in the CE CONQUAS Manual. In the case of large complex projects where different categories are involved, the weight assigned to each category will base on its value to the whole contract sum. Different CE CONQUAS will then be applied to the relevant work concerned. The total score for a project is based on the scores achieved in all assessments throughout the whole construction period. The total score will be moderated by a formula based on the contractor's performance report, which is prepared by the project engineer. The report will include consideration of the contractor's site management, compliance with statutory regulations, progress of works, safety records and public inconvenience.

5.4 PASS in Hong Kong

The demand for housing and requirement for quality have always been experienced by the Hong Kong construction industry. It has, on many occasions, been remarked that the mass production of housing units was achieved at the expense of quality construction. However, since the 1980s, discovery of poor quality housing construction has been uncommon in Hong Kong.

In order to measure contractors' performance against the defined standards and to provide a fair means of comparing the performance of individual contractors, the Hong Kong Housing Authority (HA) adopted the Performance Assessment Scoring System (PASS) on 16 February 1990. The development of PASS was based on Singapore's CONQUAS of 1989, as mentioned earlier.

Since 1991, the performance of building contractors for the new housing works has been measured on a monthly basis. The overall performance reports were also used to set the preferential tendering opportunity for the contractors, thus giving an incentive to achieve higher quality. Later, the incentive was increased by adopting the Preferential Tender Award System where contractors with higher PASS scores would get preference in the award of tenders. The 1990 version of PASS measured only the construction outputs (work produced) of contractors. Subsequent versions (1994 and 1997) of the system were developed to measure management inputs (matters relating to contractor's organizational input and management capability) as well. Furthermore, it is essential to examine the building after its occupation and assess the performance of the contractor during the maintenance period. The maintenance period assessment has also become a part of the overall performance in a contract. The most recent version, PASS – January 2002 Edition, is discussed below.

5.4.1 PASS – 2002 Edition

In March 1998, the PASS Control Unit (PCU) was formed to carry out an in-depth review of PASS. An improved system was developed under the title 'PASS 2000' and eventually promulgated as 'PASS – January 2002 Edition' (Hong Kong Housing Authority 2002).

The main assessment criteria of this version are presented in Table 5.1. The "output assessment" and "input assessment" contribute 70% and 30% respectively towards the final PASS score of the project, while "maintenance period assessment" is not counted. External Works Assessment in the previous versions of PASS has been incorporated under structural works and architectural works assessment in the present version.

Table 5.1 Overview of PASS assessment

Assessment	Wt.	Persons Assessing and the Assessment Frequency
Output Assessment		
Structural Works Assessment	70%	Monthly assessment – two months in a quarter by IT and the remaining month by PT
Architectural Works (Interim)Assessment		
Architectural Works (Final) Assessment		Once by IT after Substantial Completion Certificate is issued
Input Assessment		
Management Input Assessment	10%	Quarterly Assessed by PT
Safety Assessment	8%	
Programme and Progress Assessment	2%	
Other Obligations Assessment	10%	
Maintenance Period Assessment	—	Quarterly Assessed by PT

Wt. – Relative Weighting;   IT – Independent Team of Assessors;   PT – Project Team Members

PASS assessments are conducted through site inspections, desk-top assessments and record checks by relevant Project Team (PT) members and Independent Teams (IT). The Output Assessments of Structural Works and Architectural Works (Interim) are carried out monthly while the Input Assessment and the Maintenance Period Assessment are carried out quarterly (Table 5.1).

A frequent concern with the assessing methodology in the previous PASS versions was that the assessments were carried out solely by the members of the project team. Since the contractor performances will eventually be compared across the HA projects, there were concerns about the consistency of the assessments made by different project teams. In the meantime, Singapore had a system of independent assessors that was considered beneficial by many. Today, the responsibilities of conducting PASS assessment are shared both by Independent Teams of Assessors (IT) from the PASS Control Unit of HA headquarters and Project Team Members (PT) as shown in Table 5.1. Additionally, the Chief Architect or the Project Manager could exercise their professional judgement. At all times, the Contractor's Authorized Representative (CAR) will be invited to witness all assessments.

The factors considered in each of these measurements, and the methods of scoring are discussed in the following sections. Because a significant amount of time is required to make in-depth measurements as required by PASS, it is not practical to conduct assessments on the entire site. Hence, for all site measurements that are required to be made, the PASS manual presents methods, requirements and regulations in great detail for the selection of:

- Assessment locations through random sampling procedures
- Spots within each location
- Backup locations, in case the assessors encounter some difficulty in conducting the assessment at the original location sampled

An important PASS requirement is that, under any circumstances, no advance notice with regard to the randomly sampled assessment locations should be given to the contractor, in order to prevent them from carrying out reworking/rectification work to get a better score. After the assessment, however, the contractors will be directed to make good any defects found.

5.4.2 Output Assessment – Structural Works Assessment

The Structural Works Assessment is based on the four main factors shown in Table 5.2. These factors are subdivided into some 33 sub-factors to assess various details of structural works, e.g., size, number and shape of steel reinforcements concrete cover, and so on of buildings and external works in the housing projects.

Table 5.2 Structural works assessment

Factor		Highest Achievable Score
SW1	Reinforcement	10
SW2	Formwork and Falsework	10
SW3	Finished Concrete	10
SW4	Construction Quality and Practice	10

The assessment of factors SW1, SW2 and SW3 is based on observations made onsite, at 9 spots in 3 locations (3 spots per location) selected by sampling. The factor SW4 is based on test reports prepared by independent testing laboratories and other directives issued. Whenever assessments are made by project team members, the Project Structural Engineer or the Project Resident Engineer will carry out the assessments.

5.4.3 Output Assessment – Architectural Works (Interim) Assessment

The Architectural Works (Interim) Assessment is based on the 13 main factors AI-1 to AI-13, as shown in Table 5.3. These factors are subdivided into some 63 sub-factors to assess various details of architectural works of flats, public areas and external works. Most of the factors require onsite assessments of the quality achieved by key trades, while some factors take into account regular test results required under the contract to be conducted on critical items susceptible to latent defects.

The sampling procedure begins with the Project Clerk of Works entering the progress of work (not more than one day prior to the date of assessment) into the "Progress Entry Form" on the computer and uploading it onto the system. The information thus indicates the range of floors where various architectural works are available for inspection. In the residential tower blocks, the sampling would be based on floors and flats therein. For the other public areas and external areas, there is a zoning method given in the PASS manual. The computer would generate a random selection of locations and the PASS manual presents detailed instructions on selecting spots to be assessed. Whenever assessments are made by project team members, the Project Architect will carry out the assessments.

Table 5.3 Architectural works (interim) assessment

Factor		Highest Achievable Score
AI-1	Floor	5
AI-2	Internal Wall Finishes	10
AI-3	External Wall Finishes	5
AI-4	Ceiling	5
AI-5	Windows	5
AI-6	Plumbing/Drainage	5
AI-7	Components	5
AI-8	Precast Components	5
AI-9	Waterproofing	5
AI-10	Shop Front and Cladding	5
AI-11	External Works	5
AI-12	External Plumbing/Drainage	5
AI-13	Builders' Work and Test (Record Checking)	15

5.4.4 Output Assessment – Architectural Works (Final) Assessment

This is a one-off assessment carried out on site by the Independent Teams of Assessors (IT) from the PASS Control Unit, with the assistance of Project Clerk of Works. The assessment takes into account the 12 factors given in Table 5.4, that include some 41 sub-factors that consider many details of "as-built" workmanship and finishes to flats, public areas and external works.

Table 5.4 Architectural works (final) assessment

Factor		Highest Achievable Score
AF-1	Floor	15
AF-2	Internal Wall Finishes	20
AF-3	External Wall Finishes	10
AF-4	Ceiling	5
AF-5	Windows	5
AF-6	Plumbing/Drainage	5
AF-7	Components	5
AF-8	Precast Components	5
AF-9	Shop Front	5
AF-10	Roads/Emergency Access (External Works)	5
AF-11	Footpath/Pedestrian Areas (External Works)	5
AF-12	Cleanliness and Care of the Finishing Works	15
Total		**100**

Similar to the Architectural Works Interim Assessment, the sampling system randomly selects locations to be assessed based on floors and flats of the residential tower blocks and zones in the other public areas and external areas as stipulated in the PASS manual. The assessment will be carried out after the substantial completion of a project (or a section of the project where sectional completion is provided), when the Contract Manager has issued a completion certificate.

5.4.5 Input Assessment – Management Input Assessment

Management Input Assessment measures the contractor's site management capabilities against defined standards. There are 4 factors as given in Table 5.5, that contain some 68 sub-factors covering a wide range of aspects such as, site planning, material forecast, co-ordination, response to reported defect, payments to nominated subcontractors and so on. No sampling will be carried out, because these assessments are based on site documents and other evidence.

Table 5.5 Management input assessment

	Factor	Highest Achievable Score
IA-1	Management and Organisation of Works	10
IA-2	Resources	30
IA-3	Co-ordination and Control	40
IA-4	Documentation	20
	Total	**100**

5.4.6 Input Assessment – Safety Assessment

As apparent from Table 5.6, scores for SA1 and SA2 are drawn from the Housing Authority Safety Auditing System (HASAS) carried out by safety auditors, checked and verified by Occupational Safety and Health Council (OSHC). On a quarterly basis, OSHC will notify the Project Architect of the verified audit scores. Out of the highest achievable scores of 15 points each for SA1 and SA2, the same percentages as the relevant HASAS scores will be awarded to the contractor.

SA3 and SA4 consists of 13 sub-factors pertaining to site safety, that will be quarterly assessed by the project team at locations randomly sampled following the methods specified in the PASS manual.

Table 5.6 Safety assessment

	Factor	Highest Achievable Score
SA1	HASAS Score for Safety and Health Management System	15
SA2	HASAS Score for Implementation of the Safety and Health Plan	15
SA3	General Site Safety	35
SA4	Block Related Safety	35
	Total	**100**

5.4.7 Input Assessment – Programme and Progress Assessment

Programme and progress is assessed mainly by referring to the adequacy of the contractor's programme and the actual progress achievement. The three factors given in Table 5.7 are further divided into 15 sub-factors, e.g., programme, works plan, progress on superstructural works milestone dates for structural works, milestone dates for sample flats and so on.

Table 5.7 Programme and progress assessment

Factor		Highest Achievable Score
PA-1	Programming	20
PA-2	Progress Against Programme	40
PA-3	Milestone Dates	40
Total		**100**

No sampling is needed for this assessment. Assessments are carried out quarterly by the project team based on programming documents submitted by the contractor and the progress report compiled by the Clerk of Works for the month preceding the quarterly PASS assessment meeting.

5.4.8 Input Assessment – Other Obligations Assessment

The assessments will be carried out quarterly by the Project Architect or the Project Structural Engineer with the assistance of the Project Clerk of Works. The two assessment factors shown in Table 5.8 include eight sub-factors. Only one sub-factor of OO1, "Storage of Material", will be assessed in a zone of the site selected by random sampling. The other OO1 assessments, e.g., security, hoardings and fences, access and drainage, will represent the whole site. OO2 assessments are mainly carried out by checking the records for the whole site, however, onsite assessment may be carried out if the assessors consider it necessary.

Table 5.8 Other obligations assessment

Factor		Highest Achievable Score
OO1	Site Security, Access and Building Materials	60
OO2	Environment, Health and Other Provisions	40
Total		**100**

5.4.9 Maintenance Period Assessment

The objective of this assessment is to measure the contractor's performance during the 12-month maintenance period (defects liability period) following the certified completion of the construction project. The three factors shown in Table 5.9 have 77 sub-factors such as workmanship of outstanding work, safety for outstanding work, resources mobilized and provision of attendance for outstanding work.

Table 5.9 Maintenance period assessment

Factor	Highest Achievable Score
MPA1 Outstanding Works	40
MPA2 Execution of Works of Repair	40
MPA3 Management Response and Documentation	20
Total	100

Sampling is not carried out, thus entire site is assessed and relevant documents are considered.

5.4.10 Scoring Systems

Two types of scoring is used to arrive at the PASS score. Where HA is not ready to consider partial compliance, the assessments are decided by identifying whether there is full compliance (earning a √ on the score sheet) or non-compliance (earning a ✗ on the score sheet). Where partial compliance can be considered, the assessments are made by giving grades A, B, C, and D or N where work is not available. Methods of assessment and calculation of the PASS score may vary somewhat from assessment factor to factor and are specified in detail in the PASS manuals. Nevertheless, typical examples of both the assessment methods are given below.

Table 5.10 Two types of PASS assessments carried out

Assessments to be Graded A, B, C and D	Assessments to be Decided by Full Compliance (√) or Non-compliance (✗)
Structural Works Assessments SW1 and SW4	Structural Works Assessments SW2 and SW3
Programme and Progress Assessment	Architectural Works Assessments (both Interim and Final)
Management Input Assessment	Safety Assessment
Maintenance Period Assessment	Other Obligations Assessment

Calculating the PASS score from full compliance (√) or non-compliance (✗) assessment:

$$\text{Score for a Factor} = \frac{\Sigma \checkmark}{\Sigma \checkmark + \Sigma \times + a\, \Sigma x_c} \times \text{Highest achievable score for the Factor}$$

Where: $\sum\checkmark$ – Total number of full compliances
$\sum\times$ – Total number of non-compliances
$\sum\times c$ – Total number of non-compliances on critical items identified in the PASS manuals
a – A factor specified in the PASS manuals where a > 1 is used as an extra penalty for critical non-compliances

Assessment grades A, B, C, D and N may be given typically for:

A – Compliance
B – Minor non-compliance
C – Repeated minor non-compliance since last assessment
D – Major non-compliance
N – Work not available for assessment

From such grades, the PASS score for an assessment factor may be calculated as follows:

$$\text{Score for a Factor} = \frac{3\sum A + 2\sum B + \sum C}{3\sum(A + B + C + D)} \times \text{Highest achievable score for the Factor}$$

5.5 Other Measurement Systems

The Hong Kong Housing Authority conducted four other assessment systems: Laboratory Assessment Scoring System (LASS), Building Services Performance Assessment Scoring System (BSPASS), Maintenance Assessment Scoring System (MASS) and Building Services Maintenance Assessment Scoring System (BSMASS). The latter three are briefly described here.

5.5.1 Building Services PASS

Building Services PASS (BSPASS) is used to assess building services contractors against defined standards. BSPASS assessments comprise:

1. Output assessment is composed of
   (a) Works assessment
   (b) Substantial completion assessment

2. General (input) assessment is composed of
   (a) Assessment on management, resources and documentation
   (b) Programme and progress assessments
   (c) Safety assessment

3. Maintenance period assessment

5.5.2 Building Works MASS

Building Works MASS assessments are carried out for all building works maintenance contracts (except demolition contracts and piling contracts). MASS comprises two assessments:

1. Works assessment
2. Management assessment

The Work Assessment is conducted by HA site staff through regular on-site inspections covering three areas:

1. General obligations on site
2. Quality of works
3. Service delivery and progress of works

The Management Assessment is carried out quarterly to assess the quality and effectiveness of contractors' management input and the standards of services delivered, covering the following aspects:

1. Service to tenants
2. Response to instructions
3. Co-ordination and control of works
4. Debris / Dumping point management and Hoarding maintenance
5. Site safety
6. Competence of supervisory staff
7. Progress of works
8. Adequacy and standard of labour
9. Adequacy and standard of plant and material
10. Timely and quality submission of required documents and information
11. Timely and quality submission of cost data

5.5.3 Building Services MASS

Building Services (BSMASS) assessment is conducted for the Building Services Term Maintenance Contracts:

- Lift
- Water supply and fire services
- Electrical
- Air conditioning and ventilation
- Room cooler and split-type air conditioner (supply and installation)
- Security

The BSMASS assessment comprises the following benchmarking factors:

- Breakdown rate
- Attendance and response to emergency calls
- Routine inspection
- Breakdown repair time
- Minor repairs
- Major work
- Adherence to maintenance programme
- Submission of document
- Staff assessment
- Tenants' assessment – by way of comments and feedback

References

Bates, R., (1993), *Quality in action*, Record and Overview of Proceedings of the Conference on Quality Management in Government, Senior Staff Course Centre of Hong Kong Government, February 1993, pp.13-24.

Cheung, Y. K. Fiona, (2000), "Productivity and quality in the construction industry in Hong Kong", unpublished BSc Dissertation, The Hong Kong Polytechnic University.

Hong Kong Housing Department, (2002), *PASS Manual*, January 2002.

Kam, C. W. and Tang, S. L., (1997), "Development and Implementation of Quality Assurance in Public Construction Works in Singapore and Hong Kong", *International Journal of Quality and Reliability Management*, Vol. 14, No. 9, pp 909-928.

Lam, S.W., Low, C.M., Teng, W.A. and CIDB Singapore. (1994). *ISO 9000 in Construction*. McGraw-Hill, Singapore.

6 QUALITY AUDITS

6.1 Introduction

In Chapter 5, quality indicators, such as PASS, were discussed. Quality indicators are used to assess the quality of finished products. Quality audits, which will be discussed in this chapter, however, are used to assess the quality management systems (QMS) of construction organizations. Figure 6.1 shows the relationship.

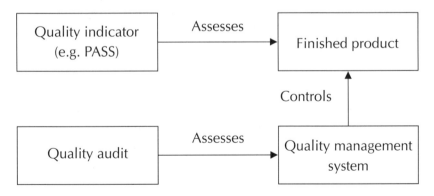

Figure 6.1 Relationship among PASS, audit, QMS and finished product

Quality audit is therefore a management process to confirm and evaluate activities related to a quality management system. As soon as a quality management system has been developed and implemented, audits should be carried out to validate the system and identify the unsatisfactory activities. This chapter highlights the importance and the scope of quality audits.

6.1.1 What Is a Quality Audit?

BS 4778 / ISO8402 defines audit as "a systematic and independent examination to determine whether quality activities and related results comply with planned arrangements and whether these arrangements are implemented effectively and are suitable to achieve objectives".

The examination can be a kind of assessment on the activities performance and available records.

6.1.2 Why Is a Quality Audit Needed?

The development and implementation of a quality assurance scheme requires testing and justification. Audit is a tool to determine the effectiveness of the quality scheme and expose the inadequacy of the system for improvement. Many standards stipulate that a quality system needs to be audited by third parties who are independent of the company providing the scheme.

6.1.3 Who Audits?

An auditor — a person or party undertaking the audit — can come from the same company, other independent organizations, government agencies, or even employed by the client who is interested in the products or services provided by the company.

Qualified auditors should have undergone an approved auditing training and have adequate experience in audit. They should also have experience in the particular type of work being examined or assessed. The personality of the auditors is also important; during cross examination and questioning, they should be patient, polite, tactful and trustworthy. This is to minimize misunderstanding and create a good friendly atmosphere in carrying out the audit.

6.2 Types of Quality Audits

Generally, audits can be categorized into three types:

6.2.1 Internal Audit Carried out by the First Party

Organizations perform these audits according to their own procedures and systems. Normally the auditors are from the same organization but coming from other departments. Experts from outside may be appointed if their expertise and experience is relevant in some important assessments.

6.2.2 External Audit Performed by the Second Party

Before making a major decision in purchasing agreement, many clients may demand an audit of the quality systems of the product or service suppliers. These audits may be performed by the in-house staff of the client or agencies specially appointed by the client for the purpose.

6.2.3 Audits Undertaken by Third Parties

Usually, the service or product providers will have their quality systems audited by third parties, such as HKQAA, which are independent bodies specialized in quality system certification. They are mainly accredited certification bodies and perform audits against the ISO 9000 quality standard. If the audit is found to be satisfactory, the independent body will award a certificate to the company under audit. This is the so called ISO 9000 certification.

Sometimes a pre-audit may be carried out before the audit. This is to ensure that the company is well prepared and equipped before the major third party certification audit.

6.3 Audit Procedures

A quality audit usually consists of the following activities:

6.3.1 Preparation Stage

As soon as the parties agree on the performance of an audit, the first action by the auditors would be contacting the auditee. The audit programme is to be agreed, the objective of the audit will be clearly stated, and relevant information will be requested. The auditor should ascertain that the documents would be adequate in both contents and scope for the audit and the kind of activities that would be assessed. The auditor also draws up a plan about the time and the kind of staff to be interviewed.

The following is the two-day programme for an audit on a construction company.

First Day

am Opening meeting
    Briefing by company staff on their operations and company tour
    Quality system review:
    •   Management responsibility.
    •   Quality system.
    •   Contract review.
    •   Document and data control.
    •   Internal quality audits.
    •   Purchasing.
    •   Control of customer supplied product.
    •   Corrective and preventive action.
    •   Control of quality records.

Lunch

pm Quality system review:
    •   Process control.
    •   Product identification and traceability.
    •   Inspection and testing.
    •   Inspection and test status.
    •   Control of inspection.
    •   Measuring and test equipment.
    •   Control of non-conforming product.
    •   Handling, storage, package, preservation and delivery.
    •   Training.
    •   Servicing.
    •   Statistical techniques.

End of the first day

Second Day

am Project audit
    Auditing of a specific project

Lunch

pm Auditor's discussion
    Closing meeting

    End of the two-day audit

6.3.2 Audit Stage

Normally the audit commences with an opening meeting to ensure that all concerned understand the purpose of the audit and the procedures to be followed. After the meeting the auditors should be given the overall perspective of the company or the site by a tour.

The auditors will examine the staff by interviews who should know their role, the requirements of the system and the objective evidence such as the duty list and the test record. They may ask questions or request staff members to actually perform an operation. Deficiencies will be identified and the auditees will be requested to explain and clarify any related issue.

After the interviews there will be a closing meeting for the auditors to present their findings. The management side of the company will have the opportunity to explain and clarify any misconception that may appear.

6.3.3 Report and Follow-up Stage

A report will be drafted to fulfill the purpose of the audit. For example, for an internal audit, the main purpose is to ascertain and improve the performance of the auditee. The audit findings, the auditee's response and follow-up corrective actions will be included in the report. For third party accreditation audit, the examination on implementation of corrective actions will be part of the audit and the findings will be included in the report.

The following is an example of a draft report of an audit on a construction company. Findings of implementation of corrective actions have not been included.

6.4 Audit Report

The following shows a typical audit report prepared by an auditor after a quality audit:

Opening Meeting

The opening meeting was attended by the principal staff from SAR Construction Company namely, Mr. A.B. Chan (MD), Mr. C.D. Cheung (DMD), Mrs. E.F. Wong (DMD) and other senior staff members.

Mr. Chan gave a briefing on the organization and activities of the company. The lead auditor introduced the auditing team from Quality Accreditation Organization. The purpose of the audit was explained to the principal staff and the proposed itinerary was agreed upon.

Audit

The audit was conducted both in the Head Office and Hong Kong Island South site office. Queries about the company's procedures and implementation were raised by the audit team to ascertain compliance with ISO requirements.

The discussion focused on the organization, operations and procedures adopted by the SAR Construction Company. All applicable clauses in the ISO 9002 were covered.

The auditors highlighted a number of observations during the discussion with the principal staff. The observations were summarized in this report.

Closing Meeting

The closing meeting was attended by the same staff.

The auditors summarized and explained their observations with the concurrence of SAR Construction Company.

Observations and Recommendations

A. Observations

Management responsibility

1. Scope of Company's Quality System is not defined and documented in Quality Manual.
2. Management to initial in Quality Manual to show commitment to quality.
3. Quality objectives should be cross-referenced to relevant procedures to show that they are being monitored and reviewed.
4. To update the company/site/quality organization charts. Departments and personnel mentioned in the procedures are not reflected in the organization charts.
5. Job responsibilities and authorities of personnel shown in the site and quality organization charts are not defined and documented.

6. The appendix documents have to be updated. Non-existing departments and personnel have to be deleted. Responsibility matrix to be updated to reflect/include existing departments shown in organization chart.
7. Appendix C is not cross-referenced to procedures.
8. Personnel responsible to report on various quality objectives in management representative meeting are not defined and documented. Personnel responsible for outstanding actions raised in management representative meeting are not defined and documented.
9. Management representative meetings did not address effectiveness and quality objectives of Quality System.
10. To clearly define in Quality Manual and Quality procedures if PASS score/rectification cost/accident rate are quality objectives. If the company adopts PASS as part of the company's quality objectives, it would be then necessary to keep a copy of the PASS manual on site and in office.

Quality System

11. An original set of standard forms and checklists used in each project has to be controlled and properly filed.
12. Project Quality Procedures for Island South project was not complete. Content of several sections was not available. Pages in the Procedures are to be numbered. Responsibilities and authorities defined in the Procedures should be limited to personnel shown in the site organization chart.
13. Quality Plan Attachment 1.6 and Annex I are not cross-referenced to procedures.

Contract Review

14. Company should specify the number of quotations to call. From the pre-audit, it was found that company practice was minimum one quotation.
15. There was no documentation to state that the company does not need to review unsuccessful tenders.
16. Under company's quality procedure on contract review, there was no documentation of the procedure (e.g. who, how, when) on progress claims made by the company to the consultant.

Document and data control

17. The persons responsible for preparing the documents should not also be the approving authority.
18. Under company's document and data control, the British Standards are classified as standard documents of the company. During the pre-audit, these standard documents were not found in the company. Company was also advised to keep record of statutory regulation codes and specifications which are relevant to the company's scope of works.
19. Company should look into establishing a well structured system to keep track of all incoming and outgoing mails.
20. The list of approved subcontractors/suppliers was deemed as controlled document, hence company should have the names of the authority preparing and approving, dated and revision status on the cover of this list.
21. There were no instructions or procedures as when and how to use the stamps shown in Annex C of Quality Plan, Figure 2 ("Certified true copy") and Figure 3 ("Superceded").
22. No approval status on Quality manual and Quality procedures. Document revision records for Quality Manual, Quality Procedures and Project Quality Procedures were not maintained. It is advisable to state revision status on forms and checklists.
23. Some standard forms are not numbered or titled.

Purchasing

24. In the Quality Plan, Clause 4 mentioned about in-house requirements. However, finding from pre-audit showed that the company did not possess any in-house requirements.
25. Company should set a limit to the amount claimable under petty cash and the personnel approved should be established.
26. Attachments 2.3 and 2.4 were found missing from the manual.
27. It was found that the dates of purchase were not indicated on some of the purchase order forms on site.
28. There should be a written procedure (who, how etc.) on alteration of purchase order on site.

Control of customer supplied product

29. The word "periodically" in Clause 5 of Quality Plan should be replaced by a specific time period such as biweekly.

30. various meetings held on site to control processes are not documented in Quality Procedural Manual e.g. consultant/client/contractor and contractor/subcontractors meetings.
31. Quality Plan Attachment 6.3 and 6.4 is not cross-referenced to procedures. Attachment 6.11 was not legible.
32. Island South Project Safety Management Programme was not updated. Safety checklists not suitable for site checking should be removed and replaced with current checklists.

Control of inspection, measuring and test equipment

33. Procedure of Quality Plan Attachment 3.3 is not documented in Quality Plan.
34. No procedures on control of inspection, measuring and test equipment used by the subcontractors. There is a need to ensure that the subcontractors use only calibrated equipment as required.

Inspection and test status

35. To review and improve Quality Plan table 2. To define exactly the methods used on site.
36. No procedures on inspection and test status for material samples and shop drawings.

Handling, storage, packaging, preservation and delivery

37. Material log book was not used at all on site. The equipment log form did not reflect those equipment on loan from the head office.
38. Practice on issuing material on first in first out basis was not done on site.
39. Defect checklist should be cross-referenced to work instruction manual.

Quality records

40. Training records of the staff were not found in the list of quality records. The retention periods of these training records have to be specified.
41. Retention period of all the quality records should have a base year of reference period.
42. Procedure on disposition of quality records should be documented and approving authority has to be specified.

43. Control of quality for old projects are subject to quality system audit.

Internal quality audit
44. Audit programme is not available. All departments, projects and ISO clauses should be audited at least once a year.
45. Checklists for internal quality audits are not used.
46. Control on issue of report number is not documented in Quality Procedure.
47. Procedures on investigation and record of the clause of non-conformance are not documented in Procedural manual.
48. To review and improve audit corrective action status log book.

Training
49. No procedure to define that records for trade tests are kept by labour department.

Servicing
50. Quality Plan Clause 5.2.1 procedure should address the authority approving the servicing report put up by the foreman.

Statistical techniques
51. It was found that the monitoring of concrete cube strength was not consistently done.

B. Recommendations

The current versions of Quality Assurance Manual and Quality Procedure Manual can be improved by adding in missing procedures and elaborating more on existing procedures that are too brief. Objective evidence and records need to be documented and maintained.

The auditors view that the quality systems documents have to be amended to address the observations raised in this audit. Some of the documented procedures were not well implemented and compiled with strictly e.g. internal quality audit. The company is advised to implement the revised procedures for another three to four months. It also needs to carry out at least one round of internal quality audits and management review to determine if the revised quality systems are suitable and effective before applying for the next audit.

References

Ashford, J.L. (1989). *The management of quality in construction*. E. & F. N. Spon.

Mills, C.A. (1989). *The quality audit – a management evaluation*. McGraw-Hill.

Stebbing, L. (1989). *Quality assurance – the route to efficiency and competitiveness*. second edition, Ellis Horwood Ltd.

7
TOTAL QUALITY MANAGEMENT (TQM) IN THE CONSTRUCTION INDUSTRY

Total quality management (TQM) is the ultimate target for any quality management system implementation. Implementing QA is usually a stepping stone for implementing TQM. QA implementation has been thoroughly discussed in the earlier chapters. This chapter focuses specifically on TQM theories, principles and tools and their applications. Findings from a couple of research projects have also been discussed to demonstrate the utility and application of TQM principles in the construction industry.

The primary purpose of TQM is to achieve excellence in customer satisfaction through continuous improvements of products and processes by the total involvement and dedication of each individual who is a part of that product/process (Ahmed 1993). The principles of TQM create the foundation for developing an organization's system for planning, controlling, and improving quality (Deming 1993).

In 1992, the Construction Industry Institute in Austin, Texas, published Guidelines for Implementing Total Quality Management in the Engineering and Construction Industry. They show that TQM has resulted in improved customer satisfaction, reduced cycle times, documented cost savings, and more satisfied and productive work forces (Burati and Oswald 1992). In the Contractor's Business Management Report's 1996 Contractor Management Survey, 44% of the respondents have or have had TQM programmes. One said TQM "raises awareness of quality and helps document result." Another said it's a "must if you are going to be the best in your area." And a third replied, "We continue to identify ways to improve."

Despite the above, the construction industry has generally lagged behind other industries in implementing TQM. The main reason for that was the perception that TQM was intended for manufacturing only (Chase *et al.* 1992). Another major was the notion that TQM was costly and required a long time period for implementation. One aspect of TQM that has frustrated the construction industry the most has been "measurement" (Hayden 1996).

Many construction companies in the US, Singapore, UK, and other European countries have been using TQM successfully for a number of years now and reaping rich rewards in improved client, consultant, and supplier relations, reduced "cost of quality", on time and within budget project completions, and a well informed and highly motivated team of staff.

Inspection traditionally has been one of the key attributes of a quality assurance/quality control system in the construction industry. Regarding inspection, Deming says, "Routine 100% inspection is the same thing as planning for defects — acknowledgement that the process cannot make the product correctly, or that the specifications made no sense in the first place. Quality comes not from inspection, but from improvement of the process" (Deming 1982). This does not mean that inspection is unnecessary. Instead, it means that more effort is put into preventing errors and deficiencies.

Quality management is a critical component in the successful management of construction projects (Abdul-Rahman 1997). Quality and quality systems are topics which have received increasing attention worldwide (Chan 1996a; Docker 1991; Kam and Tang 1997; Lowe and Seymour 1990; Tang and Kam 1999; Walters 1992; Yates and Aniftos 1997). Diverse management factors — including support of senior management, appropriate leadership style, cultivation of employee's enthusiasm and participation, open communication and feedback — must be managed properly to achieve quality management systems (QMS) in the construction industry. Planning, engineering design, and construction, often involve "one of a kind" projects where a QMS that emphasizes effective management practices is more appropriate.

The current move towards performance specification contracting in the engineering and construction industry has been extended into the quality assurance programs to formalize the expanded definitions of quality within the project development process. As a result, agencies are instituting strong construction and procurement oversight programs in order to assure that quality design and workmanship is provided in a timely manner.

The construction industry has been following a path that has led to lack of trust and confidence, adversarial relations, and increased arbitration and litigation. The industry has become increasingly reliant on burdensome specifications, which seldom says exactly what the owner intends them to say. This has led the owners to shift more of the risks to the contractors (Ahmed 1989). The net outcome is that the construction industry has been bogged down with paperwork, defensive posturing, and generally tends to have a hostile attitude towards the other participants. TQM can help reverse this trend. Although TQM is not a magic pill or panacea for all illnesses, it will, if properly implemented, help construction companies improve and will help all the parties come closer.

Attainment of acceptable levels of quality in the construction industry has long been a problem. Huge expenditures of time, money and resources, both human and material, are wasted each year because of inefficient or non-existent quality management procedures. TQM was firstly proposed in manufacturing industry and was adopted in the construction industry recently. Although quality control in the construction industry is parallel to that of the manufacturing industry, the quality control procedures that work effectively in a mass production industry have not been considered suitable for the construction industry, because almost all the construction projects are unique and there is no clear and uniform standard in evaluating overall construction quality. Thus, there is great potential and challenge for quality improvement in the construction industry.

7.1 Total Quality Management (TQM)

Total Quality management (TQM) is the art of managing the whole to achieve excellence (Besterfield *et al.* 2003). TQM provides the overall concept that fosters continuous improvement in an organization. The TQM philosophy stresses a systematic, integrated, consistent, organization-wide perspective involving everyone and everything. It focuses primarily on total satisfaction for both internal and external customers, within a management environment that seeks continuous improvement of all systems and processes (Samuel *et al.* 1998).

TQM emphasizes the understanding of variation, the importance of measurement, the role of the customer and the commitment and involvement of employees at all levels of an organization in pursuit of such improvement to fully satisfy agreed customer requirements (Besterfield 1994). TQM is now an integrative approach that forms part of the day-to-day business decision process.

7.2 Principles of TQM

As shown in Figure 7.1, five core principles are embodied in TQM (Harris 1995).

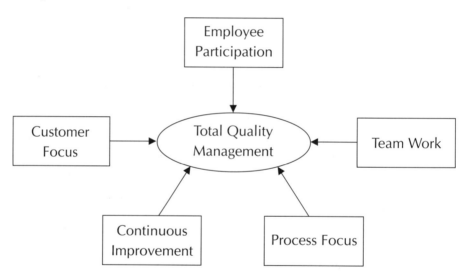

Figure 7.1 Principles of TQM

7.3 Supporting Elements

The five principles of TQM can be achieved in an organization with the aid of six basic supporting elements (Tenner and Detoro 1992):

1. *Leadership*: Senior Management must lead this effort by example, by applying the tools and language, by requiring the use of data, and by recognizing those who successfully apply the concepts of TQM.
2. *Education and Training*: Quality is based on the skills of all the employees and their understanding of what is required. Educating

and training employees provides the information they need on the mission, vision, direction, and strategy of the organization as well as the skills they need to secure quality improvement and resolve problems.
3. *Supportive Structure*: Senior managers may require support to bring about the change necessary to implement a quality strategy.
4. *Communications*: Communications in a quality environment may need to be addressed differently in order to communicate to all employees a sincere commitment to change.
5. *Reward and Recognition*: Teams and individuals who successfully apply the quality process must be recognized and suitably rewarded, so that the rest of the organization will know what is expected.
6. *Measurement*: The use of data is of paramount importance in installing a quality process. To set the stage for the use of data, external customer satisfaction must be measured to determine the extent to which customers perceive that their needs are being met.

7.4 Characteristics of the Construction Industry

Construction works are carried out in the form of projects. Projects are becoming progressively larger and more complex in terms of physical size and cost. In the modern world, the execution of a project requires the management of scarce resources: manpower, material, money, and machines to be managed throughout the life of the project, from conception to completion. The projects have five distinctive objectives to be managed: scope, organization, quality, cost and time (Figure 7.2). Construction work requires different trades and knowledge but the management, scheduling, and control of those projects utilize the same tools and techniques, and are subject to constraints of time, cost, and quality. There are also unique characteristics of projects, which differ from routine operations.

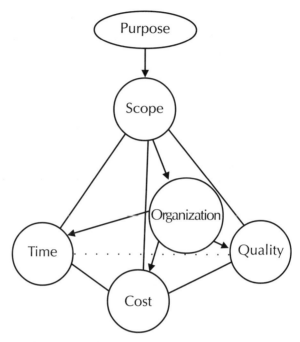

Figure 7.2 The five objectives of a project

7.5 Critical Success Factors in TQM in Construction

TQM has gained widespread global acceptance. However, some organizations have achieved remarkable success while others have suffered dismal failures. Many of the failures can be attributed to a misunderstanding of TQM or the way the organization had implemented TQM.

7. 5. 1 Customer Focus

According to the philosophy of TQM, total customer satisfaction is the goal of the entire system, and a pervasive customer focus is the means to achieve that. The function of the construction industry is to provide customers with facilities that meet their needs. For a company to remain in business, this service must be provided at a competitive cost. TQM is a management philosophy that effectively determines the needs of the customer and provides the framework, environment, and culture for meeting those needs at the lowest possible cost. By ensuring quality at each stage in the construction process, and thereby minimizing costly rework as well as other costs, the quality of the final products should satisfy the final customer.

By definition, customers may be either internal or external. The external customer is the consumer or client, in other words the end user of the products or services being offered. An internal customer is a second process or department within the organization, which depends on the product of the first. For example, for designers the products are plans and specifications, and the customers are the owner and the contractor responsible for the construction. For the contractor, the product is the completed facility, and the customer is the final user of the facility. There are also customers within the construction organization. These internal customers receive products and information from other groups of individuals within their organization. Thus, satisfying the needs of these internal customers is an essential part of the process of supplying the final external customer with a quality product.

Every party in a process has three roles: supplier, processor, and customer. Juran defined this as the "triple role concept" (Figure 7.3). These three roles are carried out at every level of the construction process. The designer is a customer of the owner. The designer produces the design and supplies plans and specifications to the contractor. Thus, the contractor is the designer's customer, who uses the designer's plan and specifications to carry out the construction process and supplies the completed facility to the owner. The owner supplies the requirements to the designer, receives the facility from the contractor, and is responsible for the facilities operation (Burati 1993). This clearly illustrates that construction is a process, and that TQM principles that have been applied to other processes are potentially adaptable to the construction industry.

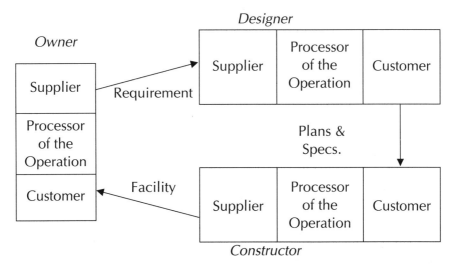

Figure 7.3 Juran's triple role concept applied to construction

7.5.2 Process Improvements

A process is a way of getting work done. A process consists of the tasks, procedures and policies necessary to satisfy an internal or external customer need (Adrian 1995). According to the TQM philosophy, if the process is correct, so will be the end result (product). Thus the organization should work to improve the process so as to improve the end product or service.

Three different approaches have emerged for improving the efficiency or effectiveness of a process. Continuous improvement is an approach used on an ongoing basis for incremental gains. Benchmarking should be used periodically, and reengineering can be launched occasionally to achieve dramatic breakthrough.

By focusing on process by measurement and analysis, a process can possibly be improved by changing the "five M's" of the process: man, machine, material, method, and measurement. A strong emphasis on process improvement centres on the measurement of variation, the control of variation, and the knowledge of variation to seek improvement. This analysis is referred to as statistical process control or statistical analysis. This is at the centre of process improvement. The objective of measuring the variation in a process is to learn how to control the variation and also how to improve the process by viewing variation as a tool for improvement. The analysis of the positive side (good performance or quality) of the variation of process is referred as a "breakthrough improvement" or "breakthrough management", which is another key component of TQM (Arditi and Gunaydin 1998).

7.5.3 Continual Improvements

The goal of continual improvement is common to many managerial theories; but TQM is unique in that it provides a specific step-by-step process to achieve this. This process consists of nine steps:

1. Identify the process.
2. Organize a multi-disciplinary team to study the process and recommend improvements.
3. Define areas where data is needed.
4. Collect data on the process.
5. Analyze the collected data and brainstorm for improvement.
6. Determine recommendations and methods of implementation.
7. Implement the recommendations outlined in step 6.
8. Collect new data on the process after the proposed changes have been

implemented to verify their effectiveness.
9. Circle back to step 5 and again analyze the data and brainstorm for further improvement.

The nine-step cycle emphasizes four elements: focusing on the progress, measuring the process, brainstorming for improvement, and verification and re-measurement. This is illustrated in Deming's Plan-Do-Check-Action (PDCA) diagram (Figure 7.4). The PDCA diagram stresses removing the root cause of problems and continually establishing and revising new standards or goals (Deming 1986).

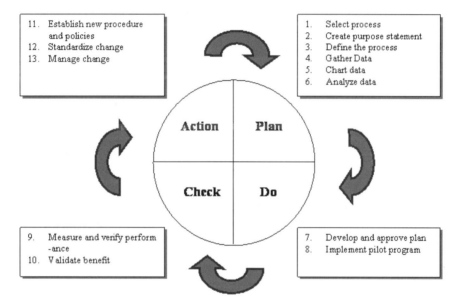

Figure 7.4 The PDCA diagram

Under TQM, management in the construction industry has two functions: (1) to maintain and improve current methods and procedures through process control, and (2) to direct efforts to achieve, through innovation, major technological advances in construction processes.

The incremental improvement of the process is achieved through process improvement and control. In every construction organization, there are major processes by which all the work is accomplished. However, there are innumerable parts in the construction process. Through the use of flow diagrams, every process can be broken down into stages. Within each stage, input changes to output, and the methods and procedures directing the change of state (i.e. the construction procedures) can be constantly improved to better satisfy the customer at the next stage. During each stage the employees should communicate closely with their supplier and customer to optimize the work process for that stage. This requires all employees to recognize their place in the process and their respective supplier and customer.

7. 6 Quality Improvement Techniques

TQM demands a process of continued improvement aimed at reducing variability. An organization wishing to support and develop such a process needs to use quality management tools and techniques. It is prudent to start with the more simple tools and techniques: check-sheet, check-list, histogram, Pareto analysis, cause-and-effect diagram (Fishbone diagram), scatter chart, and flowchart.

7. 6. 1 Check-sheet

Check-sheets are used to record events, or non-events (non–conformances). They can also include information such as the position where the event occurred and any known causes. They are usually prepared in advance and are completed by those who are carrying out the operations or monitoring their progress. The value of using check-sheet is retrospective analysis to help with problem identification and problem solving.

7. 6. 2 Check-list

Check-lists are used to tell the user if there is a certain item which must be checked. As such, it can be used in the auditing of quality assurance and to follow the steps in a particular process.

7. 6. 3 Histogram

Histograms provide graphical representations of the individual measured values in a data set according to the frequency of occurrence. It helps to visualize the distribution of data and there are several forms of histograms, which should be recognized, and in this way they reveal the amount of variation within a process. Histograms should be well designed so that staff members who carry out the operation can easily use them.

7. 6. 4 Pareto Analysis

This is a technique employed to prioritize problems so that attention is initially focused on those which have the greatest effect. It was put forward by an Italian economist, Vilfredo Pareto, who observed how the vast majority of wealth (80%) was owned by relatively few of the population (20%). As a generalized rule for considering solutions to problems, Pareto analysis aims to identify the critical 20% of causes and to solve them as a priority.

7. 6. 5 Cause-and-Effect Diagram (Fishbone Diagram)

The cause-and-effect diagram, developed by Karoa Ishikawa, is useful in breaking down the major causes of a particular problem. The shape of the diagram looks like the skeleton of a fish. This is because a process often has a multitude of tasks footing into it, any one of which may be a cause. If a problem occurs, it will have an effect on the process, so it will be necessary to consider the whole multitude of tasks when searching for a solution.

7. 6. 6 Scatter Diagram

The relationship of two variables can be plotted in a scatter diagram. A scatter diagram is easy to complete and a linear pattern obviously reveals a strong correlation.

7. 6. 7 Flowchart

A flowchart uses a set of symbols to provide a diagrammatic representation of all the steps or stages in a project process or sequence of events. A flowchart assists in documenting and describing a process so that it can be examined and improved. Analyzing the data collected on a flowchart can help to uncover irregularities and potential problem areas.

7.7 Benefits of TQM

From the viewpoint of the individual company, the strategic implications of TQM include:

- Survival in an increasingly competitive world
- Better service to its customers
- Enhancement of the organization's "shareholder value"
- Improvement of the overall quality and safety of its facilities

- Reduced project duration and costs
- Better utilization of the talents of its people

7.8 Barriers in the Implementation of TQM

Employees generally show resistance to the introduction of TQM for a host of reasons, which included fear of the unknown, perceived loss of control, personal uncertainty, the "it may mean more" syndrome, and an unwillingness to take "ownership" and be committed to change. Other barriers include (Love *et al.* 2001):

- Perceived threat to foreman and project manager roles
- Disinterest at the site level
- Lack of understanding of what TQM was, particularly on site
- Geographically dispersed sites
- Fear of job losses
- Inadequate training
- Plan not clearly defined
- Employee skepticism
- Resistance to data collection (e.g. rework costs, non-conformances material waste)

7.9 Reasons to Implement TQM

TQM efforts should be implemented to increase productivity of the organization (quantity of performance); increase quality (decrease error and defect rate); increase effectiveness of all efforts; increase efficiency (decrease time requirements while increasing productivity). Do the right things the right way!

- Quantity
- Quality
- Effectiveness
- Efficiency

7.10 The Process of Implementing TQM (TQM Concept Map, 2003)

The process of implementing TQM includes:

1. Establish as top priority by top management
   - Visionary
   - Set aggressive goals
   - "Walk the talk"

   First, the top management of the organization must establish that total quality is a top priority of the organization. Executives must provide a clear and reasonable vision; set aggressive goals for the organization and each unit, and most importantly demonstrate their commitment to TQM through their actions.

2. Cultural change
   - Paradigm change
   - Credibility development
   - Time

   Second, the culture of the organization must be changed so that everyone and every process embrace the concept of total quality management. The organization must change its paradigm to adapt to a customer-focus emphasis where everything done in the organization is aligned with exceeding customer expectations. This becomes an "on-going way of life" for the organization, continually improving and adapting. As part of the culture change, credibility among the employees must be built through rewarding positive steps toward the vision of TQM. The organization must also allow time for the change to occur and be indoctrinated into the everyday aspects of the organization.

3. Establish small teams – give overall goals
   - Define quality
   - Identify what customers want
   - Measure and change

   Third, small teams need to be developed throughout the organization to define quality, identify customer wants, and measure progress and quality. These teams will be responsible for creating their own goals, given the organization's overall goals.

4. Execute change and continuous improvement

   Finally, change and continuous improvement must be implemented, monitored, and adjusted based on analysis of the measurements.

7.11 Safety and TQM

A research was undertaken to investigate safety improvement using TQM Principles (Ahmed *et al.* 2002). Safety improvement begins with the systems-thinking approach. The systematic approach of safety management is similar to that of TQM. Hence, with the adoption of core TQM principles and procedures, existing safety management systems could be vastly improved. A simple illustration of the way in which a safety system could be evaluated and improved using TQM principles is presented in this case study. Figure 7.5 shows that different domains of a safety management system can be systematically improved using TQM principles and procedures.

As suggested by Weinstein (1997), some applications of TQM principles are helpful to occupational or work safety. According to this research, selected TQM concepts and techniques have been identified for integration with the existing safety management system. Each of these concepts is adapted to a TQM style management system in which the elements of customer focus, leadership commitment, and employee empowerment drive a continuously improving system.

Suggested TQM principles to assist the planning as well as implementation of the safety management system are illustrated in Table 7.1. Problems identified in the previous section are considered as primary factors for improvement.

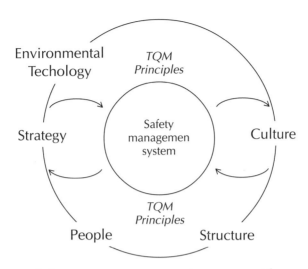

Figure 7.5 Safety management system improvement framework using TQM principles

Three major categories of TQM principles are selected:
1. *People-oriented principles*, which include customer focus, company culture, and leadership commitment.
2. *Management type*, which comprises effective communication, leadership and commitment, responsibility, benchmarking, coaching, quality planning and management.
3. *Process-control techniques*, which consist of structural problem solving and planning tools such as Pareto diagram, cause and effect diagram, affinity diagram, matrix diagram, control charts and statistical process control (Associated General Contractors of America, 1993).

7.12 Use of Quality Function Deployment in Civil Engineering Capital Project Planning

Quality function deployment (QFD) is a tool of TQM. A research was undertaken to explore the applicability of QFD in civil engineering capital project planning (Ahmed *et al.* 2003). The team proposed a QFD model that concentrated on the six basic project management areas:

- Project scope (functional requirement)
- Budget costing
- Scheduling

Table 7.1 Suggested TQM concepts and techniques to improve construction safety

Safety System (14 process elements)	Primary Problems/ Factors for Improvement	Areas to be Improved	Suggested TQM Principles
Safety and health policy	Lack of commitment, lack of discipline	Leadership, culture, commitment, employee involvement	Customer focus, company culture, Leadership commitment
Safety organization	Lack of planning and control	Leadership commitment, employee involvement	Leadership commitment, effective communication
Safety training	Lack of knowledge, lack of training	Leadership commitment, knowledge, skills	Leadership commitment, 5S*, coaching skills
In-house safety rules	Lack of planning and control, lack of training	Planning, control and review, culture	Quality planning and management
Safety inspection	Lack of inspection	Planning and control	Statistical process control
Personal protective equipment	Lack of communication and training	Planning, employee responsibility	Employee responsibility
Accidents/incidents investigation	Lack of information, lack of control	Planning and control, management by facts	Statistical process control, cause-effect diagrams
Emergency preparedness	Lack of planning and control	Planning and control	Quality planning, structural problem solving methods
Evaluation, selection and control of sub-contractors	Lack of commitment, lack of discipline	Leadership commitment, culture	Customer focus, company culture
Safety committees	Lack of commitment, lack of discipline	Leadership, culture, employee empowerment and involvement	Company culture, Leadership commitment, effective communication
Risks assessment	Lack of planning and control	Planning, control and review	Failure mode effects, analysis, fault tree analysis
Safety and health promotion	Lack of commitment and knowledge	Leadership, culture, commitment	Company culture, benchmarking
Process control program	Lack of planning, control and supervision	Planning,, control and review	Quality planning and management
Health assurance program	Lack of planning, poor knowledge	Planning control and review	Quality planning, problem solving steps
Safety resources	Lack of commitment	Leadership, commitment	Leadership commitment

- Land requirements
- Technical and safety requirements
- Statutory and environmental requirements

Data from two projects of different type, nature and scale are fed into the model for testing. Verification has given encouraging results, suggesting the validity of the QFD model. It is found that the use of the QFD can enhance the project planning process in the following ways:

1. QFD serves as a road map for navigating the planning process and always keeps track of customer requirements and satisfaction. This actually helps eliminate human inefficiency.
2. The process of building a QFD matrix can be a good communication facilitator that helps break through the communication barriers between client and the designer and among members of the design team.
3. QFD can be an excellent tool for evaluating project alternatives, balancing conflicting project requirements, and establishing measurable project performance targets.
4. QFD can be used as a quick sensitivity test when project requirements change.

Despite being preliminary and to some extent limited, the study findings are very encouraging, indicating that QFD as a project planning tool can bring benefits and enhancements to civil engineering capital project planning. Some research topics suggested for further study are streamlining the QFD process, computer-aided QFD applications, evaluation of the cost and benefits of using QFD, use of QFD in detailed design, and how to integrate QFD with total project quality management systems.

7.13 Case Study: TQM in Construction Industry in USA

A study of TQM in the construction industry in Florida was undertaken (Ahmed et al. 2002). The main objectives of the study were:
1. Investigate the adoption and implementation of TQM in the construction industry.
2. Determine the processes ("what to measure") that are most suitable and appropriate for measurement during the construction project life-cycle.
3. Develop a model ("how to measure") for the measurement and evaluation of the quality performances of the construction processes identified in 2 above as a tool for continuous improvement.

In terms of measuring work process, the construction industry does not enjoy a good reputation. The problem can be attributed to the nature of the industry, which lacks solid data gathering and the exceptional fluctuation in productivity.

7.13.1 Phase I

The first stage of this study identified the current implementation and adoption of TQM principles in the construction industry through an in-depth questionnaire. The questionnaire was divided into six parts. The conclusions from these six sections are briefly discussed below.

1. *Knowledge of TQM.* The results from this section show that the majority of the contractors agree that if a contractor satisfies its clients, the profits would increase in the long run. They feel that TQM will work very well in their organizations and the programme will be beneficial for their organizations. They are, however, not aware of any implementation programmes. Most of them feel that TQM is a philosophy used to improve cost estimating and warranty claims. This shows their lack of knowledge about TQM and the potential benefits in implementing this programme in their organizations.

2. *Perception of quality.* The majority of the contractors perceive quality as a competitive advantage next to elimination of defects. They feel that product/service quality is very important in gaining customer satisfaction because it ultimately translates into higher profits for them. They indicate that customer satisfaction is their main goal. In this section, they were also asked to rank, in the order of importance, the following attributes: Quality, Safety, Time, Cost and Scope; interestingly they ranked Scope and Cost as the important considerations, followed by Timeliness, Safety and Quality.

3. *Data acquisition method.* The results of this section show that the majority of the companies do collect data to measure the performance of operations, and that they handle problems by assigning an individual to solve them. In terms of gathering suggestions from customers, 52% of the companies have a system for doing this, but just 28% measure customer satisfaction through questionnaire surveys, and 20% gather customer suggestions via complaints or other methods. In most cases (52%), the suppliers and the subcontractor are rated, and when defects in services are identified, they are required to pay for or correct them.

4. *Quality in the organization.* Although only about 50% of the contractors surveyed have a clear definition of quality in their organizations, 86% are aware of the importance of quality. The majority of the respondents do not have a formal quality improvement programme (QIP) in place. Those that do, however, have the full support of their top management.

Also, they use a mix of quality control, TQM and ISO 9000 principles in their QIP. Demanding customers, management commitment and competitive pressures were identified as the key reasons for implementing the quality improvement programmes. The main objectives of the QIP are employee involvement followed by increasing productivity and cost reduction. The results also show that 40% of the contractors feel that the quality of their products and services improve after implementing such a programme.

5. *Training.* In the majority of the companies, employees are not given formal training in TQM or other quality improvement programmes. Only 44% of the companies report that managerial/supervisory staff has undergone quality improvement training, while 29% of the companies provided training on quality management philosophies to non-managerial and non-technical staff. Training programmes mostly emphasize customer satisfaction as a primary goal, followed by teamwork and communication.

6. *Barriers to implementing TQM.* The following is a list of the obstacles, in order of importance, to the implementation of TQM, from the most important obstacle to the least important one:

    a. Changing behaviour and attitude
    b. Lack of expertise/resources in TQM
    c. Lack of employee commitment/understanding
    d. Lack of education and training to drive the improvement process
    e. Schedule and cost treated as the main priorities
    f. Emphasis on short-term objectives
    g. Tendency to cure symptom rather than getting to the root cause of a problem
    h. Too many documents required (lack of documentation ability)

It is easy to infer from the above that although TQM has been a magic word in the construction industry for the past few years, methods and techniques to implement the quality management programme in the industry are yet to be developed. The basic reason for the lack of expertise or resources for implementing quality improvement programmes is the difficulty in assessing what to measure and how to measure them, particularly the intangible aspects of quality. Without measurement, the notion of continuous improvement is hard to follow.

To take care of the above, an attempt has been made to measure the "client satisfaction index". This provides a direct reference point from where quality improvement steps can be initiated in the construction industry. Various possibilities of client satisfaction were thus listed and rated. This is then a measure or an index of client satisfaction or dissatisfaction. This is discussed in Phase II of the study.

7.13.2 Phase II

In the second phase of this study, a second questionnaire focusing on the customer was created to identify the processes for improvement.

The analysis of the results of the second survey led to a client satisfaction index which listed the major causes of client dissatisfaction:
- Lack of attention to client priorities
- Poor planning
- Poor scheduling
- Inadequacy in processing change orders
- Poor delivery schedules and methods

Next, a third questionnaire was developed using the cause-and-effect diagram to identify the sub-causes for the main causes of client dissatisfaction. This questionnaire was presented to contractors and their feedback was sought through structured interviews. The results are summarized below.

The major sub-causes for the lack of attention to client priorities were:
- Lack of personnel training
- Lack of quality and cost control
- Inadequacy in contractor-subcontractor coordination
- Lack of conformance to specifications

The main sub-causes for poor planning in the construction industry were:
- Low quality of material and workmanship
- Poor management of change orders
- Poor cash flow analysis
- Construction underestimation
- Poor equipment management

The sub-causes for poor scheduling were:
- Incomplete design
- Lack of site condition supervision
- Inadequacy in project management coordination
- Improper network model selection

The sub-causes for the inadequacy in processing change orders were:
- Changes in design by clients
- Errors in construction design
- Defective materials/equipment
- Weather delays

The sub-causes for the poor delivery schedules and methods in the construction industry were:
- Ambiguity in methods
- Change orders from Procurement Department
- Availability of materials and equipment.

7.13.3 Phase III

In the third and final phase of the research, the most important sub-causes identified by the contractors were presented to the same clients who had participated in the second questionnaire. In the fourth questionnaire they were asked to indicate a measure of their satisfaction if the contractors made improvements to the identified sub-causes. This is called the "improvement index".

Below are the areas that would increase customer satisfaction the most if the contractors improve upon them:
- Construction underestimation
- Conformance to specifications
- Project management coordination
- Design changes by clients
- Change orders from Procurement Department

Customer satisfaction can be greatly enhanced by improving construction underestimation, conformance to specifications, project management coordination, design changes by clients and change orders from the procurement department. The above areas were identified after analyzing the results of the fourth questionnaire in which the clients were asked to identify the most important processes that need improvement.

This process can be repeated until different areas of improvement are identified through another cycle of the development of the client satisfaction index and the improvement index. The key is to understand that the client is a moving target – their expectations and requirements are constantly changing. To keep up with their ever-changing goals, contractors need to have in place a system of identifying, measuring and continuously improving their tangible and intangible products and services.

Hopefully, this study has succeeded in demonstrating how these objectives can be achieved. There is no intent on the part of the author to imply that the identified main causes and sub-causes of client satisfaction or dissatisfaction are in any way statistically significant. The objective of this study was to develop and demonstrate how a system of continuous improvement can be put in place by measuring different processes.

References

Abdul-Rahman, Hamzah (1997). "Some observations on the issues of quality cost in construction". *International Journal of Quality & Reliability Management*, 14(5), pp.464-481.

Adrian, James J. (1995). *Total Productivity and Quality Management*. Stipes Publishing, Champaign, IL.

Ahmed, Syed M. and Aoieong, Raymond T. (1998). "Analysis of Quality Management Systems in the Hong Kong Construction Industry", Proceedings of the 1st South African International Conference on Total Quality Management in Construction, Cape Town, South Africa, pp. 37-49.

Ahmed, Syed M. (1993). "An Integrated Total Quality Management (TQM) Model for the Construction Process." *Ph.D. Dissertation, School of Civil & Environmental Engineering, Georgia Institute of Technology*, Atlanta, GA, USA.

Ahmed, Syed M., Tang, P., Azhar, S. and Ahmad, I. (2002). "An Evaluation of Safety Measures in the Hong Kong Construction Industry Based on Total Quality Management Principals". *CIB-W65/W55 10th International Symposium Construction Innovation and Global Competitiveness*, Cincinnati, Ohio, USA, 6-10, September, pp.1214-1227

Ahmed, Syed M., Li, Pui Sang and Torbica, Zeljko M. (2003). "Use of Quality Function Deployment in Civil Engineering Capital Project Planning". *ASCE Journal of Construction Engineering & Management*, 129(4), pp.358-368.

Aoieong, Raymond T., Tang, S.L. and Ahmed, Syed M. (2002). "A process approach in measuring quality costs of construction projects: model development". *Construction Management and Economics*, 20(2), pp.179-192.

Arditi, David and Gunaydin, H. Murat (1998), "Factors that affect Process Quality in the Life Cycle of Building Projects", *Journal of Construction Engineering and Management*, ASCE, New York, USA, Vol. 124, No. 3, pp. 194-203.

Associated General Contractors of America (1993). *Implementing TQM in a Construction Company*. Publication No. 1211, The Associated General Contractors of America, Alexandria.

Besterfield D. H. (1994). *Quality Control*, Prentice-Hall.

Burati Jr., James L. and Oswald, Thomas H. (1992). "Implementing TQM in engineering and construction." *J. Mgmt. in Engg.*, ASCE, 9(4), 456-470.

Burati Jr., James L. and Oswald, Thomas H. (1993), "Implementing TQM in Engineering and Construction", *Journal of Management in Engineering*, ASCE, New York, USA, Vol. 9, No. 4, pp. 456-470.

Chan, A.P.C. (1996). "Quality assurance in the construction industry", *Architecture Science Review*, 39(2), pp.107-112.

Chase, Gerald W. and Federle, Mark O. (1992), "Implementation of TQM in Building Design and Construction", *Journal of Management in Engineering*, ASCE, New York, USA, Vol. 9, No. 4, pp. 329-339

Dale. H. Besterfield., Carol Besterfield, Glen H.Besterfield. and Mary Besterfield. (2003). *Total Quality Management*, Prentice Hall.

Deming, W. E. (1982). *Quality, Productivity, and competitive position.* Massachusetts Institute of Technology, Cambridge, MA.

Deming, W. E. (1986). *Out of the Crisis.* Massachusetts Institute of Technology, Cambridge, MA.

Deming, W. E. (1993). *The new economics.* Massachusetts Institute of Technology, Cambridge, MA.

Docker, P.B. (1991). "Quality assurance standard: for the building and construction industry". *South Australian Builder*, pp.15-17.

Harris C. R. (1995). "The Evolution of Quality Management: An overview of the TQM Literature". *Canadian Journal of Administrative Sciences*, 12(2), pp.95-105.

Hayden, W. M. (1996). "Connecting random acts of quality: global system standard." *J. Mgmt. in Engg.*, ASCE, 12(3), 34-44.

Ho, Samuel K. M. (1998), "Change for the Better via ISO 9000 and TQM," *Proceedings of 3rd International Conference on ISO 9000 and Total Quality Management (Theme: Change for Better)*, School of Business, Hong Kong Baptist University, Hong Kong, April 14-16, 1998, pp. 7-14

Kam, C.W. and Tang, S.L. (1997). "Development and implementation of quality assurance in public construction works in Singapore and Hong Kong". *International Journal of Quality and Reliability Management*, 14(9), pp. 909-928.

Kerzner, H. 1994, *Project Management: A Systems Approach to Planning, Scheduling and Controlling* (5th edition), (New York: Van Nostrand Reinhold).

Love P. E. D., Treloar G. J., Ngowi A. B., Faniran O. O. and Smith J. (2001). "A framework for the Implementation of TQM in Construction Organizations". http://buildnet.csir.co.za/cdcproc/docs/2nd/love_ped.pdf

Lowe, S.P. and Seymour, D. (1990). "The quality debate", *Construction Management and Economics*, 8(1), pp.13-29.

Tang, S.L. and Kam, C.W. (1999). "A survey of ISO 9001 implementation in engineering consultancies in Hong Kong". *International Journal of Quality and Reliability Management*, 16 (6), pp. 562-574.

Tenner, A. R. and DeToro, I. J. (1992). Total quality management. Reading, MA: Addison-Wesley.

TQM Concept Map (2003). http://soeweb.syr.edu/faculty/takoszal/TQM.html

Walters, W. (1992). "The road to quality - getting started", *Chartered Builder*, Chartered Institute of Building, pp.16-17.

Yates, J.K. and Aniftos, S.C. (1997). "International standards and construction", *Journal of Construction Engineering and Management*, ASCE, 123(2), pp.127-137.

8

TRANSITION FROM ISO 9000:1994 TO ISO 9000:2000 AND INTEGRATION OF QMS WITH TQM PHILOSOPHY

8.1 Introduction

As discussed in previous chapters, the ISO 9000 (1994 version) quality assurance (QA) system is a systematic approach in satisfying given requirements and providing adequate confidence. While rework, scrap, delivery, delays etc. may be minimized by the adoption of the ISO 9000:1994 system, other defects, for example, unnoticed delays, frustration, redundant internal effort, over control, manpower inefficiency and low morale, which are largely hidden, can only be exposed and cured by the adoption of total quality management (TQM) (Ahmed and Aoieong 1998). TQM was discussed in detail in Chapter 7.

Also, in Chapter 4, we have seen that in 1997, almost all contractors and consultants in Hong Kong who obtained ISO 9000:1994 certification were interested only in satisfying clients' requirement and improving companies' reputation. This attitude may not be the best way to improve quality. Nevertheless, ISO 9000:1994 (mainly for QA) may act as a stepping stone, or an integrated first step, for the implementation of TQM. The authors had done no survey from 1998 to 2001, but their subjective experience was that there had been an improving trend in general in quality management in the Hong Kong construction industry, although the rate of improvement seemed to be slow.

With the emergence of the ISO 9000:2000 version in December 2000, all construction organizations in Hong Kong certified to ISO 9000:1994 would need to prepare themselves for re-certification to ISO 9000:2000. The deadline set by the Hong Kong Government was mid-December 2003. This new version of ISO 9000 is of a more generic process-based structure with a Plan-Do-Check-Act improvement cycle (Sung and Au 1999), which assembles to a large extent prevailing international models of TQM.

In order to know how construction organizations in Hong Kong obtained recertification from ISO 9000:1994 to ISO 9000:2000 (or integrating their QMS with TQM philosophy), a questionnaire survey for and interviews with these organizations were carried out in 2002 and 2003 respectively. The details are reported in this chapter.

8.2 Key Factors for the Successful Transition from QA to TQM

Based on literature review, interaction with construction organizations and the authors' own experience, 11 key factors for the successful transition from QA to TQM were identified. A questionnaire survey conducted in Hong Kong in 2002 covered these 11 factors:

1. Clarifying the understanding of the purpose of ISO 9000:2000 and/or TQM.
2. Recognizing the real motivations for pursuing ISO 9000:2000 and/or TQM.
3. The top management's commitment to quality and the comparable extent of their senior colleagues' commitment in other functional areas.
4. The top management's pursuits of long term financial results and customer satisfaction, their vision and their enthusiasm.
5. Clarifying the top management's understanding of the ways and behaviours which they need to change to align with a ISO 9000:2000 or TQM culture.
6. The top management's willingness to undergo such personal change and development.
7. Clarifying to what degree the top management should recognize about their existing quality personnel who may need to develop new skills and behaviours.
8. Embracing leadership practices that create an environment conducive to company-wide employee involvement.

9. Recognizing the need for reward and recognition systems supportive of involvement and empowerment.
10. Embracing a behavioral orientation which treats quality as a strategic imperative.
11. Strategic practice that recognizes the value of organizational core competence and organization capability.

The target respondents for the survey were companies implementing ISO 9000:2000 or TQM programmes. There were 21 major construction contractors and consultants in Hong Kong at the time of the survey that could satisfy the above requirement. All these firms were invited to express their opinions, but only 11 out of the 21 firms responded. Experience from the 1997 surveys showed that the results obtained from contractors and consultants were very similar, so this time only one survey questionnaire was designed for use by both contractors and consultants.

The responses to the 11 key factors (for successful transition from QA to TQM) were rated on a five-point scale. The most important factor was assigned a score of 5, and the least important factor a score of 1. Figure 8.1 shows the survey results graphically.

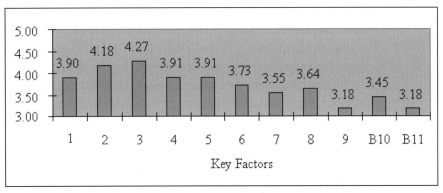

Figure 8.1 Mean score of the relative importance of the key factors for successful transition/upgrading from ISO 9000:1994 to ISO 9000:2000 and/or TQM

The above results indicated that the top four important key factors were:
- The top management's commitment to quality and the comparable extent of their senior colleagues' commitment in other functional areas. (Mean = 4.27)
- Recognizing the real motivations for pursuing ISO 9000 and/or TQM. (Mean = 4.18)

- The top management's pursuits of long term financial results and customer satisfaction, their vision and their enthusiasm. (Mean = 3.91)
- Clarifying the top management's understanding of the ways and behaviours which they need to change to align with a ISO 9000:2000 or TQM culture. (Mean = 3.91)

8.3 Important Actions for the Successful Transition from QA to TQM

Six important actions for successful transition were identified, from the authors' own experience and by referring to existing literature:

1. Compare the ISO 9000:1994 with the 2000 version and/or TQM, and develop a Quality Manual to include some new requirements of ISO 9000:2000 and/or TQM.
2. Provide sufficient training to everyone of the company including the members of the top management to ensure their quality awareness and advocate personal development.
3. Set up some tailored recognition and reward systems to promote company-wide employee involvement.
4. Set up a quality steering committee or the like whose function is to formulate the company policies and strategies on promoting quality/customer satisfaction.
5. Establish processes, such as customer satisfaction survey, to obtain and monitor information on customer satisfaction and/or dissatisfaction in order to measure the performance of the quality management system.
6. Set up quality improvement teams to pursue continual improvement to tackle inter and intra functional issues.

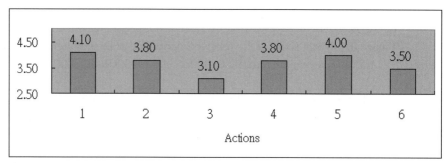

Figure 8.2 Mean score of the actions need to be taken for successful transition/upgrading from ISO 9000:1994 to ISO 9000:2000 and/or TQM

The results of the respondents' opinions (Figure 8.2) indicated that the top four important actions were:
- Compare the ISO 9000:1994 with the 2000 version and/or TQM, and develop a Quality Manual to include some new requirements of ISO 9000:2000 and/or TQM. (Mean = 4.10)
- Establish processes, such as customer satisfaction survey, to obtain and monitor information on customer satisfaction and/or dissatisfaction in order to measure the performance of the quality management system. (Mean = 4.00)
- Provide sufficient training to everyone of the company including the members of the top management to ensure their quality awareness and advocate personal development. (Mean = 3.80).
- Set up a quality steering committee or the like whose function is to formulate the company policies and strategies on promoting quality/customer satisfaction. (Mean = 3.80)

8.4 Difficulties Faced during the Transition /Upgrading Process

For this part of the questionnaire, the respondents were asked to rate on a 12-point scale. The mean score results are shown in Figure 8.3.

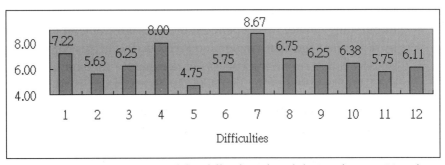

Figure 8.3 Mean scores of the difficulties faced during the transition / upgrading process

The difficulties listed in Figure 8.3 are:
1. Lack of strong senior management involvement/support.
2. Resistance or bad attitude from staff.
3. Poor internal/external communication.
4. Not fully understood by staff the requirements of ISO 9000:2000 and/or TQM.
5. Absence of well structured quality system and procedures.

6. Too much documentation and records.
7. Change in culture.
8. Insufficient quality training for staff.
9. Quality management system is not applied to sub-contractors.
10. The clients only require you to have ISO 9000 certification but they do not have quality knowledge.
11. No cooperation from the client to meet procedures under your project quality plan.
12. Maintain the ISO 9000 certification as a work permit but not really seek for quality improvement.

The results shown in Figure 8.3 indicated that the top four difficulties were:
- Change in culture. (Mean = 8.67)
- Not fully understood by staff the requirements of ISO 9000:2000 and/or TQM. (Mean = 8.00)
- Lack of strong senior management involvement/support. (Mean = 7.22)
- Insufficient quality training for staff. (Mean = 6.75)

8.5 Summary of the Survey Results

There was little argument that the quality revolution had to be led from the top. Academics and professionals were in agreement that a successful journey in pursuing quality was dependent on the degree of commitment exhibited by senior management (Taylor and Meegan 1997). This commitment would, therefore, determine the success of the quality transition/upgrading from the ISO 9000:1994 to the ISO 9000:2000 and/or the TQM.

Besides, successful transition/upgrading would also depend on recognizing the real motivation for pursing and the understanding of the purpose of the ISO 9000:2000 and/or the TQM. For example, that some construction firms used ISO 9000 as a means of producing paper work to satisfy customers was more likely to be accompanied by a motivation associated with customer pressure. Some construction firms relaxed their effort in achieving quality once their certification was obtained. For these cases, the quality systems certified were doomed to fail.

In fact, an incorrect motivation and understanding of the ISO 9000:2000 and/or the TQM might hinder an organization from sustaining its quality initiative. Transition/upgrading of quality management systems from ISO 9000:1994 to more superior standards, such as the ISO 9000:2000 or the TQM, could not proceed unless these two prerequisites were satisfied. In order to succeed, the organization should have a correct understanding of both quality concepts and the need of this transition.

The findings in the study identified the respondents' opinions in relation to the difficulties during the transition/upgrading process from the ISO 9000: 1994 to the ISO 9000:2000 and/or the TQM. The three major difficulties (out of the four shown in the last section), in descending order of importance, are discussed below.

8.5.1 Difficulty 1: Change in Culture

The biggest difficulty is the change in culture. To deal with the likely resistance to the change, two approaches should be considered: (1) employee quality awareness programme, and (2) recognition and reward system.

The employee quality awareness programme will enable the employees to know the reasons behind the change and understand the need for the change, encourage them to contribute to decision making process, and involve them in the process. It is also very important to raise the employees' awareness of quality so that the firm could survive in a highly competitive environment. When employees recognize the need for the change, they would be committed to the change. Consequently, everybody in the firm can follow the firm's objectives, and the resistance to the change can be minimized.

A tailor-made recognition and reward system might play a significant role in the achievement of company-wide employee involvement. The system should be designed to reassure and encourage employees during the period of changes and promote intrinsic motivation in order to get them excited, involved, committed, and energized.

8.5.2 Difficulty 2: Requirements Not Fully Understood by Staff

The second difficulty is that the requirements of ISO 9000:2000/TQM are not fully understood by staff. Every employee should be familiar with ISO 9000: 2000 terminology and TQM philosophy, the requirements of the quality system, and his/her role in the quality programme. It is also advisable to conduct training using a top-down approach, and to ensure that the top management understand the system first and be able to solve problems when it moved to the next level. For training to be effective, the staff's needs should be analyzed. Since not all employees require all the quality management skills, so a firm has to identify the target groups of staff and the types of training to be offered.

8.5.3 Difficulty 3: Lack of Strong Management Involvement/Support

Strong support and commitment from senior management is as crucial in the quality transition/upgrading process. It is the driving force to sustain the running of the quality programme. The top management has to generate a conducive environment in order to enhance the development of the system, and has to develop a vision and subsequently a mission statement, which should be communicated throughout the entire organization. This will enable all employees to work towards the same quality goal. The management should not just provide the necessary measures, it should also be involved actively in all phases of the transition/upgrading process.

To conclude, the transition/upgrading process does not mean leaving the ISO 9000:1994 behind to move on to something new or more superior, such as TQM or ISO 9000:2000. Rather, there were other ways in which an organization could build upon the framework provided by ISO 9000:1994 to incorporate the cultural dimensions required by the TQM or the ISO 9000:2000, in particular, (1) moving to a customer focus, (2) moving from compliance to continual improvement, and (3) moving from being reactive to being proactive.

8.6 Interviews with Construction Contractors and Consultants

Besides conducting a questionnaire survey described above, interviews were carried out with various construction contractors and consultants which were either using the year 2000 version already or in transition of the change. Companies were selected from a homepage (Hong Kong Construction Resources). The contractors and consultants as shown in Table 8.1 were interviewed in early 2003.

The shortest transition period for the change was 7 months (a contractor) and the longest time taken was 11 months (a consultant). Usually, a consultant took a slightly longer transition period than a contractor did. The following summarizes the interview results.

	Contractors	Consultants
Using Year 2000 Version	Dragages et Travaux Publics (HK) Ltd. Kin Wing Engineering Co. Ltd. Zen Pacific Construction Ltd.	Greg Wong & Associates Ltd. Scott Wilson Ltd.
Undergoing Transition	Chevalier Construction Holdings Ltd. Sun Fok Kong Construction Ltd. Dickson Construction Ltd.	Far East Consulting Engineers Ltd. Fugro (HK) Ltd. Ove Arup & Partners (HK) Ltd.

Table 8.1 Names of contracting and consulting firms interviewed

8.6.1 Difficulties Encountered in Renewing the Certification

1. The lack of trained personnel to carry out the renewal was the main problem faced.
2. In the transition, it was difficult to show that the company's quality objective is customer focused. After all, to define in detail the quality objectives of a company was already a difficult thing to do.
3. The requirement of the infrastructure (e.g. provision of computer and stationery to every staff member) is difficult to realize for some consultants.
4. It was difficult to implement the quality objectives to all levels of an organization.
5. It needed a lot of resources for converting the work description related to process approach.
6. It was difficult to make the two versions compatible and use them at the same time.

8.6.2 Significant Changes from Old to New Version

1. Customer focus was a significant change.
2. The continual improvement allowed the companies to put more emphasis on quality improvement of their products.
3. The requirements of the management committee in the new version made all levels of an organization know more about their job nature and hence facilitate the participation in achieving company's quality objectives.
4. The integrated process approach in the 2000 version was new but not difficult to comply with.

5. The new version required less redundant paper works because the documentation requirements were less stringent.
6. A significant change was the shift of the spirit of QA (quality assurance) to that of TQM (total quality management).
7. The infrastructure requirement was a significant change.
8. More meetings were to be called due to a higher focus on planning.

8.6.3 Benefits from Implementing the New Version

1. The higher emphases on management commitment and company objectives were appreciated by most interviewees as they reckoned that it was a sign of providing quality services or products.
2. Some companies (not agreed by all) could benefit from a reduction of paper work.
3. The continual improvement requirement was a good addition in the new version. This new requirement was not difficult to comply with. Companies could improve by taking corrective and preventive measures to save resources and minimize production costs.
4. The high emphasis on customer satisfaction could keep the companies on their right track of customer needs.
5. The analysis of cost data (usually for showing continual improvement) could help the companies improve their production processes.
6. An interviewee (from consultants) said both 1994 and 2000 versions could bring the same benefits to a company.
7. The new version is simpler to implement than the old version because overall speaking, there is a less emphasis on documentation procedures.

8.6.4 Demerits of Implementing the New Version

1. The collection and analysis of cost data were complicated.
2. The new version was more complicated as far as the senior management of a company is concerned. More documentation was needed for training purpose. They also had the burden to maintain the discipline at all levels and to tie all individual quality techniques to a cohesive quality system.
3. More documentation was needed for the compliance of customer satisfaction clause.
4. More meetings were to be called due to a high focus on planning.
5. A lot of resources had to be used for converting their existing working procedures to comply with the requirement of the new version. (Some interviewees did not agree with this point because their existing working procedures were quite similar to those required by the new standard.)

8.6.5 Which Version Is Better?

1. All companies (except one consultant) agreed that ISO 9000:2000 is better than ISO 9000:2000.
2. One consultant interviewee said the two versions were about the same as far as benefits to his company were concerned.

8.7 Case Studies of Companies Undergoing Transition

Two case studies on two other firms (different from the above firms) which had undergone the transition of certification from ISO 9000:1994 to ISO 9000:2000 were carried out in 2002.

8.7.1 Case Study 1: China State

China State Construction Engineering (Hong Kong) Limited (China State in short) is a subsidiary of China Overseas Holdings Limited and is one of Hong Kong's biggest and leading construction companies. It was established in Hong Kong in 1979 and had over 1,800 employees in year 2002/2003. The core business of the company involves design and construction of multi-disciplinary construction projects, such as building, civil, foundation, E&M and so on. Since December 1992, China State has received the ISO 9000 certification.

China State is committed to the mission of "honouring contracts, ensuring high quality, setting reasonable prices and abiding by righteousness" and to integrate the total quality, environmental and safety management into its long-term development strategy in order to enhance its operation and management. The information collection of this case study comprises solely a personal interview made with the Deputy Quality Manager at China State's head office.

Around July 2001, China State commenced the transition process from the ISO 9000:1994 to the ISO 9000:2000, which was carried out by the Quality and Safety (Q&S) Department of the company. In the initial stage of the transition process, the Q & S Department undertook a comparison of the existing ISO 9000:1994 quality system with the ISO 9000:2000, and accordingly revised the existing Quality Manual to address the new requirements of the ISO 9000:2000. The Deputy Quality Manager said that China State provided sufficient training to everyone in the company, including the members of the top management, to ensure quality awareness and to advocate personal development. Besides, China State had, in accordance with the requirements of the ISO 9001:2000 and under the supervision of the Quality Management Committee, formulated its year 2002 quality objectives and targets. Subsequently, China State successfully achieved the registration of the ISO 9001:2000 in January 2002.

The Deputy Quality Manager considered that the main problems faced during the transition process were "change in culture" and "not fully understood by staff the requirements of the ISO 9000:2000". Nevertheless, "lack of commitment and involvement by the top management" was not a problem during the transition process. He believed that because the company had been quite well developed, the management's decision to implement ISO 9000 quality management was firm.

The Deputy Quality Manager indicated that a change in corporate culture was the biggest difficulty for a large company to overcome. Therefore, China State regularly organized "Promotion Month" and "Seminars on Quality, Safety and Environmental Protection" to promote quality awareness of its staff in the middle of every year. Additionally, "Best Site Management Practices Forum" was also arranged at the end of every year. The purpose of this forum was to discuss the main trend, existing problems, and quality issues, as well as to set out plans for the coming year.

To incorporate the cultural dimensions required by the ISO 9000:2000, the Deputy Quality Manager said that the main effort was to concentrate on imposing quality awareness of every related parties, including company's staff, suppliers, subcontractors, and the clients, through both promotion programs and forums held in year 2002. The management of the company, department heads, and site managers were actively involved in these activities. Besides, other promotion efforts were given to promotional materials, posters, magazines and newsletters, etc, in order to enhance employee's quality awareness.

Moreover, in order to let them understand the terminology and requirements of the ISO 9000:2000, an orientation program was conducted for all employees, including members of the top management. Furthermore, any new employee would have an orientation session within three months after joining the company. The Deputy Quality Manager said that since not every employee needs all the management skills, they would identify the target groups of staff to receive different types of training. The training provided was mainly a "top-down" approach so that the top management understood the quality system in the first place. It would effectively enhance the development of the system in a conducive environment.

Based on the experience, the Deputy Quality Manager considered that three factors, (1) "the top management's commitment to quality", (2) "recognizing the real motivations for pursuing ISO 9000:2000", and (3) "embracing leadership practices which create an environment conducive to company-wide employee involvement", are the important factors for successful transition. Besides, the Deputy Quality Manager indicated that "embracing a behavioural orientation, which treats quality as a strategic imperative", is an important factor that could allow the company to survive in today's competitive environment in the construction industry.

8.7.2 Case Study 2: Fraser

Fraser Construction Company Limited is a medium size company mainly involved in the construction of geotechnical and drainage works such as minipile, slope stabilization works, and drainage repair works. In April 2001, Fraser registered itself under the ISO 9000:1994 quality management system in response to meet the needs of its customers. At the same time, it also prepared to carry out the transition process to be certified to the ISO 9000:2000. Following five months of hard work and then an external audit, Fraser received the ISO 9001:2000 certification in November 2001. An interview was then conducted with a representative from the management level to review the experience of Fraser's regarding the transition from the ISO 9000:1994 to the ISO 9000:2000.

Fraser employed a consultant, Icon Resources Limited, to provide the necessary input and enable Fraser to start in the correct direction. Fraser's representative indicated that the probable problems faced during the transition process were "necessity of strong senior management involvement or support"; "change in culture" and "quality management system not applied to sub-contractors".

The representative pointed out that strong support and commitment from senior management was crucial in advocating this quality transition and in the running of the new version of the ISO 9000. In order to generate a conducive environment for the development of the system, the strong senior management commitment/support should not only provide the necessary means, but should also be involved actively in all phases of the transition process. To do this, the representative said that the training provided was mainly a "top-down" approach, the top management being the first to understand the concept of the quality management system.

The representative also indicated that the implementation of the new version of the ISO 9000 involved the culture change of an organization. Some staff members were quite reluctant to change. To eliminate the resistance to change and impose an awareness of quality, all staff, including both head office's staff and sites' staff, had to receive an induction training on the concepts and the principles of the year 2000 version of the ISO 9000 quality management system under the guidance of Icon Resources Limited. There were also other efforts in enhancing awareness. For example, some quality promotion posters showing the company policy were displayed in both the head and site offices. In addition, the top management communicated its emphasis on quality to each employee. To do so, the company's policy, mission, quality objectives and targets were issued to all staff by internal memos. All employees knew the reasons behind the change, understood the needs for the change, and were subsequently willing to work towards the same quality goal.

Like many other construction companies in Hong Kong, the subcontractors employed by Fraser were relatively small and had lower management ability. Most of them had established a good and long-term working relationship with Fraser and were generally very co-operative. To enhance the implementation of quality management on the part of the subcontractors, the management from Fraser needed to put in efforts to support their subcontractors. Fraser therefore kept motivating the subcontractors, which were long established and competent small companies, to perform good quality works by giving them continuous businesses, so as to cement co-operative relationships with them in achieving the quality goals.

In response to the requirements of ISO 9000:2000, Fraser revised their Quality Manual accordingly and arranged "Management Review Meeting" at the end of every year which would be attended by the Managing Director, representative from management level, Administration Manager, Project Managers and Site Agents, etc. to monitor the quality performance as well as to set out the action plans for the coming year.

Finally, the representative said he believed that the first most important factor for successful transition was the top management's commitment to quality. It was because only the top management had the authority and the ability to develop any changes as well as to ensure any necessary efforts that could be consistently applied. The second most important factor was "recognizing the real motivations for pursuing ISO 9000:2000". In addition, the representative considered that the new version of the ISO 9000 was relatively suitable for and simpler for implementation by small to medium size construction companies as compared to the ISO 9000:1994, because the new version places less emphasis on documentation procedures. This agrees with the earlier interview findings.

8.8 Summary of the Interviews and the Case Studies

Eleven interviewees, including six contractors and five consultants, all of which are big and reputable firms in Hong Kong, were conducted to find the impact of the certification to the year 2000 version of the ISO 9000 quality management system to these companies. A majority of construction organizations in Hong Kong were implementing or were about to implement (at the time of the interviews in early 2003) quality management systems based on ISO 9000:2000 because it was a mandatory requirement for tendering public projects by December 2003. There was evidence that the impact was positive. The ISO 9000:2000 was becoming a sign of quality services or products. The results of the interviews showed that certification to the new standard by construction organizations was achievable without insurmountable problems. Although there were difficulties in implementing the new standard, the consensus was that the ISO 9000:2000 is a good version and is in general better than the 1994 version. The conversion process for certification to the new (2000) version, as experienced by most construction related firms in Hong Kong, was reasonably easy, though moderate difficulties existed. This was confirmed by the two case studies reported in this chapter. From the studies of China State and Fraser, it was found that to change the quality culture of a firm and to convince everybody of the necessary changes were quite difficult. It required considerable effort from the management and a top-down approach, because what was really needed was not only a certification, but rather the development of the whole organization's quality culture. It was also observed that the ISO 9000:2000 is easier to adopt by relatively smaller construction companies because less documentation procedures are required by this new version.

Acknowledgements

The authors express their sincere thanks to those involved in their research, particularly Mr. Franky K.F. Chin (China State), Mr. Ringo S.M. Yu (Fraser) and Mr. Nickie G.L. Yu (former staff of Fraser) for the case studies; Mr. Liu (Zen Pacific)), Mr. Aggendon (Dragages), Mr. Sher (Scott Wilson), Ms. Lai (Dickson), Mr. Lau (Sun Fok Kong) and Ms. Cheng (Far East) for the interviews. The help of other unnamed individuals is also gratefully acknowledged. All the above gave their valuable time and expert opinions to this study and the permission to disclose their company names in this chapter.

References

Ahmed, Syed M. and Aoieong, Raymond T. (1998). "Analysis of Quality Management System in the Hong Kong Construction Industry". *Proceedings of the 1st South African International Conference on Total Quality Management in Construction, Cape Town, South Africa*, pp. 37-49.

International Organization of Standards. (1994). *The ISO 9000 standard: year 1994 version of quality management systems.*

International Organization of Standards. (2000). *The ISO 9000 standard: year 2000 version of quality management systems.*

Man, Y.F. and Tang, S.L. (2003). "From QA (quality assurance) to TQM (total quality management for the construction industry: Hong Kong experience". *Proceedings of the Second International Conference on Construction in the Twenty-First Century (CITC-II)*, Hong Kong, 10-12 December 2003.

Sung, Edmund and Au, M.P. (1999). How TQM Practices Help Companies Align to ISO 9000 Version 2000 Standards? *Proceedings of The 4th International Conference on ISO 9000 & Total Quality Management, Hong Kong Baptist University, Hong Kong, 7-9 April 1999*, Ho, K. M. Samuel (ed.) HKBU Business School & Authors, pp.52-156.

Tang, S.L., Shum, H.T. and Man, Y.F. (2004). "The change of quality management systems from ISO 9000:1994 to ISO 9000:2000 – Hong Kong construction industry experience". *Proceedings of the 4th International Conference on Construction Project management*, Nanyang Technological University, Singapore, March 2004.

Taylor, W.A. and Meegan, S.T. (1997). "Senior executives and the ISO 9000 – TQM transition: A framework and some empirical data". *International Journal of Quality & Reliability Management*, Vol. 14, No. 7, pp.669-686.

http://kingng818.tripod.com/construction/company.html. Web of "Hong Kong Construction Resources".

9 QUALITY COST MEASUREMENT I (PREVENTION, APPRAISAL AND FAILURE COSTS MODEL)

In Chapters 1 to 8, discussions have been concentrated on the implementation of construction quality systems. In order to quantify the benefits arise from implementing quality management systems, quality must be measurable. In this and the next two chapters, the measurement of quality improvement by means of "quality cost" will be discussed.

9.1 Quality Costs

Although there are numerous tools for measuring quality, the "cost of quality", or "quality costs", is considered by Juran (1951) to be the primary one. Oberlender (2000) summarized **quality costs** as follows.

> "Quality costs consist of the cost of prevention, the cost of appraisal, and the cost of failure. Prevention costs are those resulting from quality activities used to avoid deviations or errors, while appraisal costs consist of costs incurred from quality activities used to determine whether a product, process or service conforms to established requirements. Failure costs are those resulting from not meeting the requirements and can be divided into two aspects. Internal failure costs are the costs incurred on the project site due to scrap, rework, failure analysis, re-inspection, supplier error, or price reduction due to nonconformance. External

failure costs are costs that are incurred once the project is in the hands of the client. These include costs for adjustments of complaints, repairs, handling and replacement of rejected material, workmanship, correction of errors, and litigation costs."

Through the measurement of quality costs, management can be alerted to the potential impact of poor quality on the financial performance of the company. Moreover, management can also determine the types of activities that are more beneficial in reducing quality costs. Dale and Plunkett (1991) reported that, despite the great quantity of literature discussing quality cost, there was no uniform view of what quality cost means and how it can be measured. Many researchers proposed various approaches to measuring and classifying quality costs. Reviews of quality costs literature can be found in Plunkett and Dale (1987) and Porter and Rayner (1992). A brief summary of the most widely recognized approach, the prevention-appraisal-failure (PAF) model, is given below.

- *Prevention costs*. The cost of any action taken to investigate, prevent or reduce the risk of nonconformity or defects.
- *Appraisal costs*. The cost of evaluating the achievement of quality requirements.
- *Failure costs*. The costs of nonconformity both internally (discovered before delivery to the customer, including scrap, rework, re-inspection and redesigning) and externally (discovered after delivery to the customer, including warranty costs and service calls).

Although the PAF model is universally accepted for quality costing, Porter and Rayner (1992) described some of the drawbacks of this approach as follows.

- It is difficult to decide which activities stand for prevention of quality failures since almost everything a well-managed company does has something to do with preventing quality problems.
- It is sometimes difficult to uniquely classify costs into prevention, appraisal, internal failure, or external failure costs. (Example will be given later.)
- The PAF model does not include intangible quality costs, e.g. loss of reputation.
- The key focus of TQM is on process improvement, while the PAF categorization scheme does not consider process costs.

This chapter discusses whether or not the PAF model is suitable to be applied in capturing quality costs in the construction industry, and if not, what kind of a model will be more suitable.

9.2 Construction Quality Cost Approaches Commonly Used in the Past

9.2.1 Davis' Approach

In 1987, the Construction Industry Institute Quality Management Task Force developed a Quality Performance Management System (QPMS) to track quality costs (Construction Industry Institute 1989). It is a management tool to be used for the quantitative analysis of certain quality-related aspects of design and construction by systematically collecting and classifying the costs of quality. The cost of quality is defined as the cost of correcting deviations (rework) plus the cost of quality management activities. The QPMS is based on the assumption that quality costs can be adequately tracked using 11 rework causes and 15 quality management activities. The rework causes, when coupled with four phases of a project, give a total of 40 (instead of 11 × 4 = 44) deviation categories with some illogical combinations eliminated. Fifteen quality management activities covering the main activities involved in design and construction were used to track quality management costs. Nine industrial projects were tested, yielding an average cost of rework of 12.4% of the project cost.

Although QPMS is simple and flexible, Abdul-Rahman (1995) stated that QPMS does not consider the effect of failure on time-related cost and knock-on cost, i.e. the cost to speed up work to make up for lost time. Moreover, the procedure does not show the specific source of problems, but rather, five major causes of origin, i.e. the owner, designer, vendor, transporter, and constructor, were stressed (see Table 1).

9.2.2 Abdul-Rahman's Approach

Abdul-Rahman (1993) developed a quality cost matrix to capture the cost of non-conformance during construction. The matrix lists such information as "problem category", "specific problem", "when problem was discovered", "causes of problem", "extra duration needed to correct problem", "additional cost of activity", "amount of additional time-related cost", and "any additional cost".

Two construction projects were used by Abdul-Rahman to test the model and the total cost of non-conformance incurred was 5% (1995) and 6% (1996) of the tender value. In Abdul-Rahman's cost matrix, the focus was on capturing the cost of non-conformance and no attempts were made to quantify other classifications of quality cost, including prevention and appraisal costs. In problem categorization, the items used were rather broad. In order to capture the cumulative effect of problems in a specific area, effective cost coding, similar to the one used by Davis, should be employed. Moreover, Abdul-Rahman did not consider the origin of deviations as Davis did. It is the authors' view that in order to have a better understanding of the quality costs data, the question "who caused the deviation?" should be essential.

9.2.3 Low and Yeo's Approach

In view of the shortcomings of Davis' and Abdul-Rahman's approaches, Low and Yeo (1998) proposed a construction quality cost quantifying system (CQCQS). The cost system is basically a documentation matrix that accounts for quality costs expressed as prevention, appraisal and failure costs. The headings of the matrix are "cost code", "work concerned", "causes", "problem areas", "time expended", "cost incurred" and "site record reference". The main feature of this model is the use of coding to classify the various items under the heading "work concerned". The items that constitute the heading "work concerned" can be obtained from the Works Breakdown Structure (WBS) of a typical project. Under the heading of "cost code", various components of quality costs, such as prevention, appraisal and failure costs can be classified. However, similar to Abdul-Rahman's approach, the matrix was designed to capture the cost of failure primarily, and little effort was made to quantify the prevention and appraisal costs. Moreover, Low and Yeo also did not consider the origin of deviations.

9.2.4 Other Approaches

Barber *et al.* (2000) developed a method to measure costs of quality failures in two major road projects. The method was largely based upon work-shadowing. Personnel on site were shadowed for a period of time and the quality problems encountered were recorded. Only direct costs of rework for the failures and the related costs of delay were included in their study. Of the weekly budgeted cost of the specific areas of work studied, failure costs amounted to about 16% and 23 % of the weekly budget respectively.

Love and Li (2000) quantified the causes, magnitude and costs of rework experienced in two construction projects. Data were collected from the date on which construction commenced on-site to the end of the defects liability period. A variety of sources such as interviews, observations, and site documents were used to collect the rework data and only the direct costs of rework were included. The costs of extra work, including variations and defects, were found to be 3.15% and 2.40% for the two projects. It was also concluded that the primary causes of rework were changes initiated by the client and end-user, together with errors and omissions in contract documents.

9.2.5 Summary of the Three Major Approaches

1. Measurement of failure cost was the focus of the first three models. However, only Abdul-Rahman and Low and Yeo pinpointed the causes of the problem. While Abdul-Rahman used wording to describe the problem area, Low and Yeo found using codes to be more convenient.
2. Measurement of prevention and appraisal costs is important due to the fact that it could be a significant part of the total quality costs. Without quantification, the relationship between prevention cost, appraisal cost, and failure cost cannot be fully understood. However, the definition of what constitutes quality management activities is somewhat arbitrary and therefore an exact definition of each activity is necessary.
3. Extensive coding was used in Davis' model to capture quality management costs and deviation costs. However, when compared with Low and Yeo's model, the latter is easier to implement and less costly.
4. Only Abdul-Rahman considered the effect of failure on time-related cost and knock-on cost. The heading "any other additional cost" was used in his cost matrix to capture any other remedial costs incurred indirectly due to the failure of an activity. It is essential to include the above as part of the cost of failure because this cost could be substantial. Without measurement, the full impact of the failures on the project cost cannot be determined.
5. Due to the short time frame involved, Low and Yeo's model was not tested.
6. The specific sources of deviations were not considered in Davis' approach. It is essential to pinpoint the specific source of deviations so that preventive measures can be employed and improvements can be made.

The three models' similarities and differences are summarized in Table 9.1.

Table 9.1 Summary of three models designed to capture quality costs

	Measurement of Prevention & Appraisal Costs	Measurement of Failure Cost	Origin of Deviation*	Specific Source of Deviation**	Use of Coding System	Effect of Failure	Field Tests
Davis	√	√	√	×	√	×	√
Abdul-Rahman	×	√	×	√	×	√	√
Low & Yeo	×	√	×	√	√	×	×

* Origin of deviation: owner, designer, vendor, transporter, constructor.
** Specific source of deviation: formwork, reinforcement, concrete, etc.

9.3 A Suggested PAF Quality Cost Model

In principle, applying the quality cost concept to the construction industry is straightforward. However, in practice, it is rather complex. A recent survey carried out in Hong Kong (Ahmed and Aoieong 1998) indicates that close to 60% of the respondents do not measure costs related to defects. The main reason for that could be just a lack of interest in knowing the quality costs. Moreover, it is rather difficult to measure quality costs without the implementation of an effective quality cost tracking system. Last but not the least, all personnel, from top management to site staff, should be aware of the usefulness of quality cost data to the company. The top management should also do their best to remove any negative views on the system. Some of the main features of an "ideal" quality tracking system are listed below.

1. The quality tracking system should be able to capture all components of quality costs. With one or more components missing, the effect of varying one component on the others cannot be visualized. Within each component of quality costs, standardization of category is necessary so that meaningful data can be compared between projects. On the other hand, the categorization should be flexible enough to ensure that the system can be modified when necessary to suit different types of projects.

2. The use of coding system in tracking quality costs is essential. As suggested by Low and Yeo (1998), the coding system should be compatible to the local quality assessment systems, if any.

3. The ease of use of the quality cost tracking system is essential. The system must be simple to use because the people who would be collecting cost data are the personnel on site. Any extra workload created from the

system must be kept to a minimum. Due to the highly competitive environment in the construction industry, it is impractical to implement any extra system that would result in much extra workload to site staff.

4. The practicality of the quality cost tracking system is essential. The main barrier to trying out the system is the long construction time involved. Recent interviews with some project managers also indicated the following barriers:

- Most companies considered quality costs data confidential and were therefore very reluctant to release any data. The top management must be convinced that collecting quality costs is beneficial to the company.
- Lack of resources is another barrier. Depending on the complexity of the system, extra personnel would be involved in implementing the system and that will result in extra cost to the company.

9.4 Improvement Made Based on the Existing Models

In PAF models, the three components of the quality cost in building contractor's perspective are prevention cost, appraisal cost, and deviation (failure) cost. In order to obtain a general view from the industry on the categorization of quality costs, extensive in-depth interviews with contractors and local authorities should be conducted as a first step in developing the model. A list of quality management activities used in construction projects and a list of common deviations that occurred on these construction sites can then be established. Other than costs incurred from the quality-related activities and deviations, the origin of deviations is also a concern as a deviation by any of the participants may cause quality costs for that particular participant or for others. Typical quality management activities, cause and source of deviations are shown below.

Prevention Activities
Activities used to avoid deviations or errors:
- Quality system development
- Quality program development
- Personnel training
- Specifications/Design review

Appraisal Activities
Activities used to determine whether a product or process conforms to established requirements:

- Materials inspections/tests
- Inspection
- Maintenance of testing/measuring equipment

Causes of Deviations

The reasons for a deviation, which is a departure from established requirements to occur:
- Communications errors
- Defective materials
- Design errors or omissions
- Poor workmanship
- Faulty equipment

Origin of Deviations

Davis' QPTS (1987) had identified five origins: owner, designer, vendor, transportation and constructor.

Specific Source of Deviations

Specific area where problems occur, for example, formwork, reinforcement, concrete … etc.

9.5 Development of the Coding System

In order to record and trace quality costs effectively, every item under each component of quality costs, the origin of deviations, and the source of deviations should be codified. The procedures can be summarized as follows:

- Conduct an investigation of the general practices in the construction industry with regard to the use of coding system.
- Investigate the accounting procedures of contractors and attempt to integrate the coding system with the accounting cost coding system.
- Typical Task Codes can be generated from the Work Breakdown Structure (WBS) of a typical project using the standardized divisional breakdown (Masterformat) developed by the Construction Specification Institute (CSI). A master list of section titles and numbers can be found in the appendix of Oberlender's book (2000). Major "divisions" include *site work, concrete, metals, wood and plastics, doors and windows, conveying systems, mechanical and electrical*, etc. An example of the master list is shown in Table 2.

- After the development of the categorization of quality costs as discussed above, "management activity code", "task code" and "cause code" can then be generated. The origin of deviations can also be traced by using a proper coding system.

Table 9.2 Typical master list developed by the Construction Specification Institute

Section Number	Title
DIVISION 3 — CONCRETE	
03100	**CONCRETE FORMWORK**
-110	Structural Cast-in-Place Concrete Formwork
-120	Architectural Cast-in-Place Concrete Formwork
-130	Permanent Forms
03200	**CONCRETE REINFORCEMENT**
-210	Reinforcing Steel
•	
•	
•	
03700	**CONCRETE RESTORATION AND CLEANING**
-710	Concrete Cleaning
-720	Concrete Resurfacing
-730	Concrete Rehabilitation
03800	**MASS CONCRETE**

9.6 Development of the Quality Cost Model

With the categorization of quality costs developed and the coding system in place, the next phase is the development of a quality cost system to capture all the quality-related costs.

In the development of the system, the users who would be operating the system must first be identified. Most likely, the person who would be analyzing quality costs data will be the quality manager of the project. A user-friendly computer program developed to track quality costs will definitely make the system easy to implement. In the development of the computer program, flexibility is essential to ensure that the system can be modified when necessary to suit different types of projects. On the other hand, the people who would be recording quality costs data will be the site staff. Considering their heavy workload, attention must be given to the design of the site quality cost record sheets so that staff resistance can be reduced to a minimum. Typical site record sheets for recording "prevention and appraisal" and failure costs are shown in Figures 9.1 and 9.2 respectively.

Figure 9.1 Sample of prevention and appraisal costs record sheet

Name *Chan Tai Man*
Project *XYZ*

1	2	3	4	5	6	7
Date	Task Code	Prevention Activity	Appraisal Activity	Time required	Cost incurred	Remark
Feb. 1	03110	P4		1 hr.	500.00	Formwork review
Feb. 2	03210		A3	2 hrs.	800.00	Reinforcement Inspection
Feb. 3	03310		A2	3 hrs.	1200.00	Concrete Testing
Feb. 5	03210		A3	1 hr.	400.00	Reinforcement Inspection
Feb. 6	03310		A2	3 hrs.	1200.00	Concrete Testing

Prevention Activity Code	Activity
P1	Quality system development
P2	Quality program development
P3	Personnel training
P4	Specifications/Design review
•	•
•	•
•	•

Appraisal Activity Code	Activity
A1	Maintenance of equipment
A2	Materials inspections/tests
A3	Site inspection
•	•
•	•
•	•

Figure 9.2a Sample of failure cost record sheet for Division 3

  Name   <u>*Lee Tai Man*</u>
  Project   <u>*XYZ*</u>
  Division 3 - CONCRETE

1	2	3	4	5	6	7	8
Date	Task Code	Cause Code	Origin Code	Extra Time*	Cost incurred	Additional Cost**	Remark
Feb. 10	03110	D4	C	5 hrs.	2500.00	7000.00	Poor workmanship in formwork
Feb. 13	03250	D3	D	4 hrs.	2000.00		Missing anchors in beams
Feb. 15	03350	D2	V	10 hrs.	5000.00	2000.00	Wrong type of floor finishes

In recording the failure cost, record sheets can be grouped together according to divisions (or sub-divisions) as shown in Figure 9.2a and 9.2b. At the end of a project, with quality costs data collected from all sources, reports relating task, prevention, appraisal and failure costs can be easily generated from the program as shown in Figure 9.3. Depending on the information required by the end-user, quality costs reports can be presented in different formats. By grouping entries according to task code (Division) as shown in Figure 9.4, the prevention, appraisal and failure costs of each division can be identified. Similarly, by grouping entries according to activity code or cause code as shown in Figure 9.5, the prevention/appraisal cost of any particular quality management activity and the failure cost of a particular cause can be determined. Similarly, by grouping entries according to origin code as shown in Figure 9.6, the distribution of failure cost among the parties involved in the project can be determined.

Figure 9.2b Sample of failure cost record sheet for Division 5

Name *Ho Tai Man*
Project *XYZ*
Division 5 - METALS

1 Date	2 Task Code	3 Cause Code	4 Origin Code	5 Extra Time*	6 Cost incurred	7 Additional Cost**	8 Remark
Feb. 14	05310	D1	T	8 hrs.		2000.00	Wrong sizes of steel delivered
Feb. 25	05310	D3	D	3 hrs.	1800.00		Wrong openings in steel deck

Cause Code	Deviation	Origin Code	Origin
D1	Communication errors	O	Owner
D2	Defective materials	V	Vendor
D3	Design error	C	Constructor
D4	Poor workmanship	S	Subcontractor
•	•	T	Transportation
•	•	D	Designer
•	•		

* Extra time required for remedial work
** Additional cost incurred as a result of the failure

Quality cost measurement I (prevention, appraisal and failure costs model)

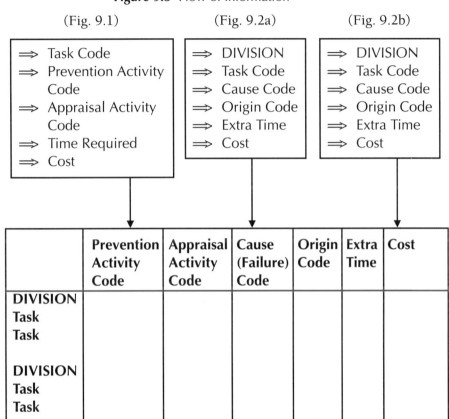

Figure 9.3 Flow of information

Figure 9.4 Sample of quality cost report sheet (sorted by tasks)

Name   _Ho Tai Man_
Project _XYZ_

	Prevention Cost	Appraisal Cost	Activity Code	Failure Cost	Cause Code	Origin Code	Extra Time	Total
DIVISION 3								
Task 03110	500.00		P4					
Task 03110				9500.00	D4	C	5 hrs.	
Task 03210		800.00	A3					
Task 03210		400.00	A3					
Task 03250				2000.00	D3	D	4 hrs.	
Task 03310		1200.00	A2					
Task 03310		1200.00	A2					
Task 03350				7000.00	D2	V	10 hrs.	
Sub-Total	500.00	3600.00	→	18500.00	→	→		22100.00 ↓
DIVISION 5								↓
Task 05310				2000.00	D1	T		↓
Task 05310				1800.00	D3	D		↓
Sub-Total				3800.00	→	→		3800.00 ↓
TOTAL								**25900.00**

Figure 9.5 Sample of quality cost report sheet (sorted by prevention/appraisal/failure)

Name   _Ho Tai Man_
Project _XYZ_

–	Task Code	Prevention Cost	Appraisal Cost	–	Task Code	Failure Cost
Activity Code				**Cause Code**		
A2	03310		1200.00	D1	05310	2000.00
	03310		1200.00	Sub-Total		2000.00
Sub-Total			2400.00			
				D2	03350	7000.00
A3	03210		800.00	Sub-Total		7000.00
	03210		400.00			
Sub-Total			1200.00	D3	03250	2000.00
					05310	1800.00
P4	03110	500.00		Sub-Total		3800.00
Sub-Total		500.00				
				D4	03110	9500.00
				Sub-Total		9500.00
TOTAL		**500.00**	**3600.00**			**22300.00**

Figure 9.6 Sample of quality cost report sheet (sorted by origins)

Name _Ho Tai Man_
Project _XYZ_

	Task Code	Failure Cost
Origin of Deviation		
O (Owner)	—	—
D (Designer)	03250	2000.00
	05310	1800.00
V (Vendor)	03350	7000.00
C (Contractor)	03310	9500.00
T (Transportation)	05310	2000.00
TOTAL		22300.00

9.7 Feedback from the Industry about the Suggested PAF Model

In-depth interviews with 12 leading construction contractors in Hong Kong were conducted between 1998 and 1999. The details of the PAF approach described above were presented to the interviewees. Comments regarding the implementation of such a quality cost tracking system based on the PAF approach were obtained and summarized in the following paragraphs.

9.7.1 Interviews

The purpose of the interviews was to obtain views from personnel involved in quality assurance on issues relating to quality cost in the construction industry. A total of 12 contractors were selected from the Hong Kong Housing Authority list of Building Contractors. They represented 38% of all the contractors who were eligible to tender for contracts valued at above HK$300M (HK$7.80 = US$1.00). Nine interviewees were quality managers while the remaining three held senior managerial positions.

9.7.2 Discussion on Results of Interviews

Awareness of Quality Cost in the Construction Industry

All contractors interviewed have obtained ISO 9000 certification. Although all expressed that they were aware of the concept of quality cost, there was no uniform view of what quality costs were and what should be included under the quality cost umbrella. When the interviewees were asked if they knew what was meant by quality costing, the majority thought that it was just a process to measure the cost of non-conformance. The reason for this incomplete view of quality costing could be traced to the nature of their jobs. As most of the interviewees are quality managers, their main concern is probably the cost related to non-conformances.

As far as the classification of quality cost is concerned, the majority was aware of the quality cost terms of prevention, appraisal, and failure. However, after understanding the correct definition of those terms, most of the interviewees agreed that, in order to avoid confusion, standardized classification of quality cost for the construction industry was essential.

Quality Cost Measurement

None of the interviewees measured quality costs according to the PAF classification. Their reasons could be summarized as follows:

1. **No system set up for the measurement.**

    In order to capture the quality costs of a project effectively, a quantitative system is essential. No such system has been established in any one of the companies interviewed. The reason for not establishing such a system is due to the lack of resources. One interviewee also commented: "The design of such a system is not too difficult. However, it is nearly impossible to implement such a system on a construction project site."

2. **No resources available for the measurement.**

    Whether or not a company has extra resources to implement a quality cost measurement system depends very much on its profit margin. All interviewees shared the view that measurement of quality costs is beneficial. However, due to the highly competitive nature of the construction industry, most contractors have to cut their profit margin to a minimum so as to win a tender. For this reason most contractors are reluctant to spend more to improve their quality system. One quality manager commented: "If we commit more on quality, more than the minimum required level, our construction cost will be higher than our competitors and therefore the project will most likely be awarded to others."

3. **Multi-level contracting hinders the measurement of the total quality costs.**

    Subletting works by the main contractor to subcontractors is a very common practice in Hong Kong. Over 90% of the labourers in a building site belong to subcontractors (Tang *et al.* 2003). It is also very common to have a few levels of contracting between the general contractor and the laborers. In order to measure the quality costs of a particular trade, it would require the full cooperation of all levels of subcontractors. Bearing in mind that the subcontractors are under no obligation to do additional work to collect quality cost data, it would be rather difficult to obtain their cooperation even though there is a strong commitment from the general contractor. In order to capture the total quality costs of a project, all sub-contractors from different trades have to be involved.

From the general contractor's point of view, most contractors interviewed were not enthusiastic about obtaining facts on quality costs. This is simply because most of the projects are contracted out and only the final "product" and not the "process" is their concern. From the subcontractors' point of view, instead of committing extra resources to collect data (which hinders the effectiveness in processing the works), they would much rather allocate their resources to rectify problems than recording and analyzing quality costs. Moreover, there is a negative side of the system because contractors/subcontractors will have to admit and report their mistakes during the implementation of the system, which they would rather not do.

Benefits of Measuring Quality Cost

In spite of the negative feedback, all interviewees shared the view that there were merits in measuring quality costs. Some benefits are listed as follows:
- Measuring quality costs helps in identifying problem areas.
- Measuring quality costs helps in reducing the cost of non-conformances.
- Results from quality cost analysis can be used in the contractor/sub-contractor selection process.
- Quality cost measurement is an effective tool for the implementation of TQM.

Personnel Responsible for Measuring Quality Cost

Most interviewees indicated that quantity surveyors and site engineers would be in the best position to record quality costs on site. They are the persons on site who are responsible for the assessment and inspection of work done.

9.7.3 Interview Results

In addition to the above comments, all interviewees expressed that such a suggested model would be too complicated to use in practice and straight implementation of the PAF model might not be possible. It was probably due to the complexity of the structure of the construction industry. The interviewees also suggested that a more practical model, if feasible, to measure quality costs should be developed for the benefits of the construction industry.

However, it may be **feasible** that the PAF model is applied to **small construction projects**. If a project is really small, the resources required to capture the PAF costs are not that large and the number of levels of subcontracting for the project can be quite small. In such circumstances, the PAF model for capturing quality costs can be practicable. Hall and Tomkins (2001) reported a successful case using the PAF model to capture the quality costs of a small construction project in the UK. The authors of this book believe that the suggested PAF model in this chapter is good for small construction projects to capture PAF quality costs.

References

Abdul-Rahman, H. (1993). "Capturing the Cost of Quality Failures in Civil Engineering". *International Journal of Quality & Reliability Management*, Vol. 10, No. 3, pp.20-32.

Abdul-Rahman, H. (1995). "The cost of non-conformance during a highway project: a case study". *Construction Management and Economics*, Vol. 13, No. 1, pp.23-32.

Abdul-Rahman, H. (1996). "Capturing the cost of non-conformance on construction sites", *International Journal of Quality & Reliability Management*, Vol. 13, No. 1, pp.48-60.

Ahmed, Syed M. and Aoieong, Raymond T. (1998). "Analysis of Quality Management Systems in the Hong Kong Construction Industry". *Proceedings of the 1st South African International Conference on Total Quality Management in Construction*, Cape Town, South Africa, pp.37-49.

Aoieong, Raymond T., Tang, S.L. and Ahmed, Syed M. (2002). "A process approach in measuring quality costs of construction projects: model development". *Construction Management and Economics*, Vol. 20, No. 2, pp.179-192.

Barber, P., Graves A., Hall, M., Sheath, D. and Tomkins, C. (2000). "Quality failure costs in civil engineering projects". *International Journal of Quality & Reliability Management*, Vol. 17, No. 4/5, pp.479-492.

Construction Industry Institute (1989) "Costs of Quality Deviations in Design and Construction". *Quality Management Task Force Publication 10-1.*

Dale, B.G. and Plunkett, J.J. (1991). *"Quality Costing"*. Chapman & Hall.

Davis, K. (1987). *"Measuring Design and Construction Quality Costs"*. Source Document 30, Construction Industry Institute.

Hall, M. and Tomkins, C. (2001). "A cost of quality analysis of a building project: towards a complete methodology for design and build". *Construction Management and Economics*, Vol. 19, No. 5, pp.727-740.

Juran, J.M. (1951). *"Quality Control Handbook."* First Edition, McGraw-Hill Inc.

Kuprenas, J.A., Soriano, C.J., and Ramhorst, S. (1996). "Total quality management implementation and results". *Practice Periodical on Struct. Design and Constr.*, ASCE Vol. 1, No. 2, pp.74-78.

Love, P.E.D. and Li, H. (2000). "Quantifying the causes and costs of rework in construction". *Construction Management and Economics*, Vol. 18, pp.479-490.

Low, S.P. and Yeo, K.C. (1998). "A construction quality costs quantifying system for the building industry", *International Journal of Quality & Reliability Management*, Vol. 15, No. 3, pp.329-349.

Oberlender, G.D. (2000). *Project Management for Engineering and Construction*. McGraw-Hill, Inc. pp.324.

Porter, L.J. and Rayner, P. (1992). "Quality Costing for Total Quality Management". *International Journal of Production Economics*, Vol. 27, pp. 69-81.

Plunkett, J.J. and Dale, B.G. (1987). "A review of the literature on quality related costs". *International Journal of Quality & Reliability Management*, Vol. 4, No. 1, pp.40-52.

Tang, S.L., Poon, S.W., Ahmed, S.M., and Wong, K.W. (2003). *Modern Construction Project Management*. 2nd edition, Hong Kong University Press, Hong Kong.

10
QUALITY COST MEASUREMENT II (PROCESS COST MODEL)

10.1 Introduction

Unlike the production line in the manufacturing industry, the construction process is far more complicated. Due to the vast number of parties involved and the uniqueness of each activity in a construction project, straight application of the concept of quality cost based on a manufacturing setting is rather difficult. We have seen from Chapter 9 that construction professionals are in general skeptical about the practicability of PAF quality cost models for the construction industry. However, if the measurement of quality cost is beneficial to the industry, attempt should be made to design a measuring system, which is applicable to and acceptable by the industry.

Among the prevention, appraisal and failure costs, failure cost is the most difficult to identify and collect. As one quality manager expressed in an interview: "One will never get the whole picture of quality costs for a building project because there are so many parties involved." Although this comment may be right, one can always focus on some areas of construction and determine the quality costs of the work done by a particular sub-contractor. As another quality manager pointed out in an interview: "In order to easily attempt a trial on this quality cost exercise, we suggest to simplify the experimental model of building construction project in the areas where quality non-conformances are often found." In order words, if there are not enough resources to implement the system in full scale, a smaller scale implementation of the system can still be achieved. In fact, this is totally in line with the "process cost model" approach described in the BS 6143 Part 1 (1992).

In the past, the concept of quality costing has suggested that certain identifiable costs are in some way related to the "quality" of the end product. By contrast, with the current TQM culture, all activities are related to processes and therefore a cost model should reflect the total costs of each process.

In BS 6143 Part 1 (1992), some terms are defined as follows:

Process cost: The total costs of cost of conformance (COC) and cost of nonconformance (CONC) for a particular process.

Cost of conformance (COC): The intrinsic cost of providing products or services to declared standards by a given, specified process in a fully effective manner.

Cost of non-conformance (CONC): The cost of wasted time, materials and resources associated with a process in the receipt, production, dispatch and correction of unsatisfactory goods and services.

In the traditional prevention, appraisal and failure approach, many costs can be attributed to either prevention or appraisal. Design reviews, for example, may be considered to be a prevention cost; however, they are essentially a checking stage and, as such, could be considered an appraisal cost. Moreover, it is always difficult to decide which activities should be categorized as prevention cost because it can be argued that everything a well managed organization does is directed at preventing quality problems. This is one of the drawbacks of the PAF approach. A new approach, the process cost approach, presents a much simpler categorization than the PAF scheme. In the context of building construction, process cost models can be developed for any selected construction processes. The COC and/or CONC can then be measured and key areas for process improvement identified. Since both COC and CONC can usually be improved, the whole process is in tune with the fundamental goal of TQM — continual improvement. Since the TQM philosophy focuses on process improvement, the PAF approach has its limitations, as process costs are not considered in the cost categorization scheme.

10.2 The Process Cost Model for Construction Processes

10.2.1 Identification of Process

TQM requires the management of processes, not just of outputs. In order to improve the quality and productivity of a process, top management must first identify specific processes with discrete activities that require improvement. Continual improvement of processes should be established as an organization's objective. Through systematic management review, top management can easily evaluate the effectiveness of processes and actions can be taken accordingly. For example, in building construction, the process "concreting" can be isolated and selected for process improvement.

10.2.2 Defining the Process and Its Boundaries

Once a particular process is isolated, its boundaries must be properly defined so that all key activities will be included for investigation. An organization's resource availability must be considered when determining the process boundaries so that no excessive resources will be demanded. For example, the process "concreting" is bounded by the activities "Formwork construction" and "Formwork striking".

10.2.3 Identification of Inputs, Outputs, Controls and Resources

In the context of building construction processes, the elements of the process (Figure 10.1) could be expressed as follows:

Inputs: Construction materials, such as concrete, admixtures, reinforcement, steel, waterstops. . . etc., that are transformed by the construction process to create building elements/system. (They are shown at the left hand side of Figure 10.1.)

Outputs: The end products of the construction process which include:
1. that which conforms to the requirement; e.g. structural members, HVAC system;
2. that which does not conform to the requirement; e.g. defects;
3. waste; e.g. material waste;
4. construction process information; e.g. inspection reports.

(They are shown at the right hand side of Figure 10.1.)

Controls: Inputs that define, regulate and/or influence the construction process.
Examples of controls embrace construction procedures, method statements,

work plan, drawings, standards and specification. (They are shown at the top of Figure 10.1.)

Resources: Contributing factors that are not transformed to become an output. Examples of resources include labour, equipment and overhead. (They are shown at the bottom of Figure 10.1.)

Figure 10.1 Typical process model of construction processes

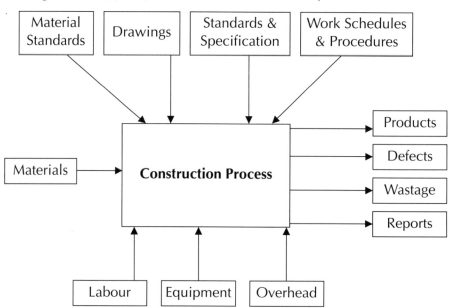

A typical process model developed based on the above concept for the process of "concreting" is shown in Figure 10.2.

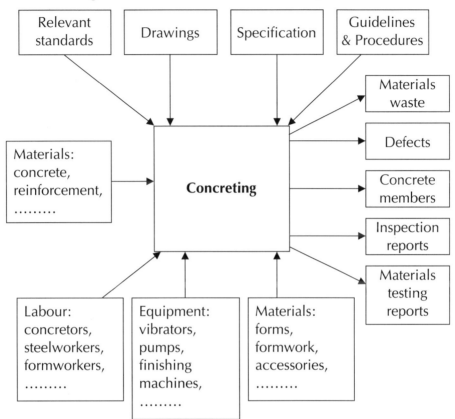

Figure 10.2 Process model for "concreting"

10.2.4 Identification of the Process Cost Elements and Calculation of the Quality Costs

Each process contains a number of key activities. Once the process model is prepared, all the cost elements associated with the key activities of the process can be identified and established as a cost of conformance (COC) and/or a cost of non-conformance (CONC). Both types of costs offer opportunities for improvement. The total process cost of the process is the sum of COC and CONC. The process owners from whom data are obtained should also be identified. In listing the activities, the contractor's coding system can also be employed so that data can be retrieved directly from the accounting system. A typical table which contains the key activities, code numbers, COC and CONC of the process "concreting" is shown in Table 10.1. This table will be used as a reference in preparing the "process cost report".

Table 10.1 Identification of costs for key activities

Process Name: Concreting			
Key activities	**Code**	**Cost of conformance**	**Cost of non-conformance**
Formwork	3110	Cost of materials and labour	Cost of waste materials
		Cost of temporary shoring	Cost of rework due to wrong dimensions
Reinforcement	3210	Cost of materials and labour	Cost of rework due to incorrect size of reinforcement
			Cost of extra labour and materials due to changes
Concrete accessories:	3250	Cost of materials and labour	Not applicable
Cast-in-place concrete	3300	Cost of materials (concrete, admixtures …… etc.) and labour	Cost of waste materials
		Cost of equipment for concrete placing	Cost of extra labour and materials due to changes
Materials testing	3301	Cost of testing of materials for conformance to standards	Not applicable
Concrete curing	3370	Cost of materials and labour	Not applicable
Rework	3201	Not applicable	Cost of extra labour and materials due to defects

10.2.5 Constructing a "Process Cost Report"

A typical process cost report is shown in Table 10.2. The report should contain a complete list of the costs of conformance and non-conformance elements. Items in Table 10.1 can be extracted directly for use in Table 10.2 when they are applicable. The report should also specify whether actual or estimated costs are used. In addition, the process owner who is responsible for a particular key activity can also be included in the report.

Table 10.2 Typical process cost report for the process "concrete"

Process cost report Process name: Concreting Boundary: "Formwork construction" to "Formwork striking" 5th floor to 10th floor of Building XXX Process owner: various					
Process Conformance	Cost Act. Est.	Process Owner	Process Non-conformance	Cost Act. Est.	Process Owner
Activity: 3110 (Formwork erection) • labour • materials		Formwork subcontractor	Activity: 3110		
Activity: 3210 (Reinforcement) • labour • materials		Steel subcontractor	Activity: 3210 (Reinforcement) • waste • rework due to wrong length		Steel subcontractor
Activity: 3300 (Concrete placing) • labour • materials • equipment		Concrete subcontractor	Activity: 3300 (Concrete placing) • rework due to honeycombs in concrete • rework due to wrong opening location		Concrete subcontractor
Total process conformance cost			Total process nonconformance cost		

10.2.6 Improvement Process

Once set up, the model is used for regular reporting on performance. Since construction processes are dynamic in nature, it is essential that the data collection process commences only after the construction processes become stable. Comparison with previous periods can then be made and areas for improvement identified. Failure costs, in particular, should be prioritized for improvement through reduction in costs of non-conformance. An excessive cost of conformance may suggest the need for process redesign. It is also essential that the process owners be involved in the improvement team. A flowchart that depicts the steps in implementing the process cost model is shown in Figure 10.3.

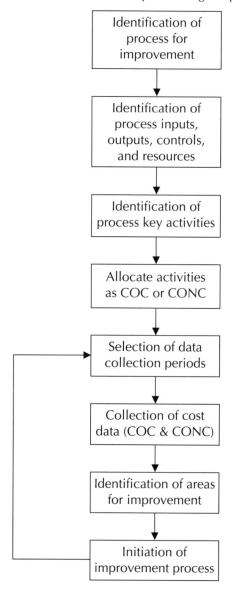

Figure 10.3 Flow chart of implementing the process cost model

A table highlighting the differences between the PAF models used by previous researchers (see Chapter 9) and the proposed process cost model is shown in Table 10.3.

Table 10.3 Differences between the models used by previous researchers and the authors

	Previous researchers' model			The authors' model
	Davis	Abdul-Rahman	Low & Yeo	
Approach	Prevention Appraisal Failure	Prevention Appraisal Failure	Prevention Appraisal Failure	Process Cost
Cost Measured	Prevention Appraisal Failure	Failure	Failure	Costs of conformance & nonconformance
Scope	Whole project	Whole project	Whole project	Selected processes
Types of project used in testing	Industrial projects	Civil engineering projects	Not yet tested	Being tested in building & civil engineering projects
Emphasis	Reduction of failure costs	Reduction of failure costs	Reduction of failure costs	Continual process improvement

10.3 Advantages of Adopting the Process Cost Model in the Construction Industry

The advantages of the process cost model can be summarized as follows:

1. The focus is no longer on capturing the "total cost" of quality of an entire project, which is rather difficult to do. Instead, specific processes in a project can be identified for monitoring and improvement. Winchell and Bolton (1987) did suggest that a small-scale PAF model could be used to measure quality costs in departments or sections of production. In the traditional PAF approach, total quality cost is essentially a build up of many cost elements in many departments. The micro approach proposed by Winchell and Bolton, on the other hand, broke down the whole system and evaluated the effectiveness of an individual department. Although Gibson et al. (1991) had tested the micro approach in a manufacturing company, no literature can be found so far on testing

such an approach in the construction industry. It is the authors' view that the implementation of the micro approach in the construction industry is much less feasible than in the manufacturing arena for the following reason: production in a manufacturing company relies on contributions from different departments within a company, whereas the "final product" of a construction company depends on the work contributed by individual subcontractors (many different companies). Also, for a PAF model, it is rather difficult to distribute accurately the prevention costs to individual construction processes, for example, the cost of maintaining a quality assurance system can be easily estimated for a construction project; however, the allocation of this cost to particular processes is not easy. Moreover, the micro PAF model does not include process costs and therefore is not fully compatible with the TQM concept of continual improvement.

2. Focus is no longer on the categorization of costs (P, A and F), which may be quite difficult and unsatisfactory. As pointed out in BS 6143 (1992): "Many of the costs can be justified as fitting into any one of the three categories, i.e. prevention, appraisal, and failure. Process cost model offers a simpler categorization."

3. Since the focus is no longer on capturing the "total cost" of quality of an entire project, the demand of resources is much lesser and much of it can be allocated to a particular process identified for monitoring and improvement.

10.4 Survey Feedback on Process Cost Model

Two questionnaire surveys were conducted in 2002, one in the US and the other in Hong Kong. In the US survey, the questionnaire was sent to approximately 550 members in the directory of the Construction Management Association of America; a total of 43 firms responded, yielding a response rate of about 8%. In the Hong Kong survey, the questionnaire was sent to 163 construction-related firms in Hong Kong; a total of 34 firms responded, yielding a response rate of 21%. The respondents were presented with the process model (Figure 10.2 of this chapter) and asked if the model was feasible and practicable for construction processes.

The majority of the respondents (79% in the US and 74% in HK) indicated that the model was feasible and only 12% in the US and 21% in Hong Kong were "not sure". Among the few respondents who considered the model as not feasible, two responded with explanations. One respondent believed that the model could only be applied to the manufacturing industry while the other indicated that the model was not applicable because he was from an engineering consulting firm. The explanations provided were an indication of their incomplete understanding of the process cost model. In fact, the process cost model can be applied to any processes or services including management processes such as the design process. In terms of the practicality of applying the model in construction processes, 30% of the respondents in the US felt that the model were practicable. The responses to this question in Hong Kong were more positive: 58% of respondents felt that the model was practicable. The reason for this could be that most of the respondents' firms in Hong Kong were certified to ISO 9000 and therefore familiar with the concept of process improvement. Most respondents indicated that the lack of time and resources to implement was the main reason for the model to be impractical. There were, however, positive comments too. According to one respondent, the COC and CONC could be easily defined. Another respondent said that the model was practicable because a process's boundaries were flexible and this allowed the management to define the boundaries according to their own situations. Table 10.4 summarizes the responses.

Table 10.4 Feasibility and practicality of the process cost model

	% of respondents			
	Feasibility		Practicality	
	US	HK	US	HK
YES	79%	74%	30%	58%
NO	9%	6%	26%	6%
NOT SURE	12%	21%	44%	35%

The respondents were then asked to select the best person(s) to provide data related to COC and CONC. Due to differences in local practices, the title "quantity surveyor (QS)" was added to the selection list in the questionnaire for Hong Kong. For COC, project manager and site superintendent were selected by the US respondents as the best persons while the HK respondents chose project manager and QS. Similar results were also obtained for CONC with increased percentages in the categories of owner and site inspector. Because the costs of non-conformance are sensitive data, the increased involvement of owner and inspector (owner's representative) is to be expected. The responses are summarized in Table 10.5.

Table 10.5 Best person(s) to provide Cost of conformance and Cost of non-conformance data

	% of respondents			
	Cost of conformance		Cost of non-conformance	
	US	HK	US	HK
Owner	17%	18%	29%	21%
Project manager	76%	61%	74%	55%
Site superintendent	52%	27%	57%	24%
Site inspector	12%	39%	19%	36%
Quantity surveyor	N/A	55%	N/A	61%
Prime Contractor	45%	48%	48%	42%
Subcontractor	36%	24%	29%	18%
Others	14%	9%	10%	3%

When the respondents were asked if they would consider applying this concept in their future projects, a big difference was observed from the two groups. Only 24% of respondents in the US indicated positively, but the figure was almost doubled in Hong Kong. This result clearly showed that the firms in Hong Kong were more open to new concepts and were more willing to pursue continual improvement in construction processes. The responses to this question are shown in Table 10.6. Typical reasons for choosing "No" and "Not sure" were "too costly", "don't know how" and "not useful" as shown in Table 10.7. Other reasons were mostly resources related. One respondent in Hong Kong, a senior quality manager, replied that the management would not believe in this concept. Another respondent in Hong Kong, a QS representing a client, pointed out that it was unlikely that the contractors and consultants would provide such cost data. In the US, however, close to 50% of the respondents who checked "No" and "Not sure" explained that the nature of their works were mainly construction management and therefore thought that the process cost model was not directly applicable. This is in fact a misconception because the model can be applied to construction management process, as has been mentioned earlier. Indeed, the process cost model can be applied to site operations as well as management processes.

Table 10.6 Application of the process cost model in future projects

	% of respondents	
	US	HK
YES	24%	47%
NO	26%	22%
NOT SURE	50%	31%

Table 10.7 Reasons for 'No' and 'Not sure' to future application of the process cost model

	% of respondents	
	US	HK
TOO COSTLY	22%	13%
DON'T KNOW HOW	19%	25%
NOT USEFUL	19%	19%
OTHERS	47%	50%

10.5 Feedback from Interviews with Construction Professionals

To obtain more information regarding quality measurements in construction firms, further in-depth interviews with construction professionals were conducted in both the US and Hong Kong in 2002. A total of 15 (6 in the US and 9 in Hong Kong) interviews were conducted. The details of the interviewees are shown in Table 10.8.

Table 10.8 Profile of interviewees' organizations and positions

Firm	Field of work	Location	Interviewees' position
1	General contractor, Construction Managers	US	Senior Project Manager
2	General contractor	US	Project Manager
3	Subcontractor	US	President
4	Engineering, Architecture & Construction	US	Principal
5	General contractor	US	Vice President
6	General contractor with ISO 9000 certified	US	Quality Control Manager
7	General contractor with ISO 9000 certified	HK	Group Quality Systems Manager
8	General contractor with ISO 9000 certified	HK	Senior Quantity Surveyor
9	General contractor with ISO 9000 certified	HK	Quality Manager
10	General contractor with ISO 9000 certified	HK	Quantity Surveying Manager
11	General contractor with ISO 9000 certified	HK	Senior Quality Assurance Manager, Registered Lead Assessor
12	Quality Assurance Agency	HK	Registered Lead Assessor
13	Certification Agency	HK	Registered Lead Auditor
14	Government Development Department	HK	Chief Engineer
15	General contractor with ISO 9000 certified	HK	Site Agent

Out of the six interviewees in the US, five indicated that the PCM was feasible for application to construction processes because of its simplicity and flexibility. However, three of them were not sure about the practicality of the PCM for the following reasons:

1. Since such a cost measurement system has never been tested and implemented before, it is quite difficult to foresee whether money will be saved after the implementation.

2. The resources involved, particularly the site staff, will be tremendous if an accurate figure of CONC is required.
3. "Fire fighting" is still a better method because quick decisions can be made. Moreover, there are other tools available and to stick to just one method may be too rigid.
4. Since tight control is usually exercised on site, no significant defect cost is anticipated in a project and therefore the process cost model might not be useful. On the other hand, if the defect cost of a project is significant, its origin can be easily traced.

Since most of the reasons given above have been traced to the fact that PCM is only a proposed concept, it is understandable why it has led to skepticism. Other reasons mentioned are all related to problems of resources. Nevertheless, three interviewees expressed their enthusiasm in applying the concept to their own construction projects in the future. One interviewee pointed out that quality measurement was not essential because defects could be reduced to a minimum level if subcontractors were carefully selected. He further believed that subcontractors should be warned that their contracts could be terminated by reason of serious and repetitive defects. Another interviewee indicated that PCM was not practical because its implementation would divert his attention from concentrating on preventive measures. Moreover, the reduction of defect costs was not his priority because of its insignificance. The quality manager from an ISO certified company, however, believed that PCM was both feasible and practical. He felt that non-conformity items on site were not too difficult to capture if site personnel such as inspectors, foremen and quality control staff made an effort to keep the records. Some of them kept these types of records anyway. This quality manager also expressed his belief that the application of this concept to construction industry was innovative and agreed that the US construction industry in general had to do more in order to be competitive. However, the implementation of such a system had to be initiated by a top management committed to quality.

In the interviews conducted in Hong Kong, all interviewees shared the view that applying the PCM model in the construction industry was much more feasible than applying the traditional prevention-appraisal-failure (PAF) model. They further indicated that when applying the PCM model, the resource level required in quality costs measurement would be more flexible due to the fact that the number of processes selected for monitoring could vary according to the resources available. Moreover, when applying the PCM model, the timing for quality costs measurement was less rigid because performance was only measured at selected periods. Three interviewees, who were quality managers, expressed their willingness to apply the concept to their construction projects. However, they also pointed out that this initiative had to be supported by the top management as well as project managers and other site staff. Two interviewees, who were quantity surveyors, believed that the collection and estimation of the cost of non-conformance would not impose any significant extra burden on the site staff. Moreover the cost of conformance could be easily extracted from the bill of quantities. Nevertheless, a few interviewees were still skeptical about the practicality of PCM giving similar reasons as those listed above (1 to 4).

In the PAF model, to capture the total quality costs of a project would require the full cooperation of all subcontractors. All interviewees agreed that, unlike the PAF model, the number of subcontractors involved in implementing PCM could be kept to a minimum, depending on the processes selected. For example, the number of subcontractors involved in the concreting process could be as low as three. This will make the collection of cost data easier. In addition, the two interviewees from certification bodies (certification officers) in Hong Kong believed that PCM could be used as a simple tool to measure process improvement, which is a requirement as stipulated in the year 2000 edition of the ISO 9000. The latter emphasizes process approach besides client satisfaction and continual improvement (Aoieong and Tang 2002).

Applications of PCM to the construction industry will be discussed in the next chapter.

References

Aoieong Raymond T., Tang, S.L. and Ahmed, Syed M. (2002). "A process approach in measuring quality costs of construction projects: model development". *Construction Management and Economics*, Vol. 20, No. 2, pp.179-192.

Aoieong Raymond T. and Tang, S.L. (2002). "The Year 2000 Version of ISO 9000 and the Process Cost Model for measuring Quality Costs in Construction Processes". *Proceedings of the first International Conference on Construction in the 21st Century*, Editors: Irtishad Ahmad, Syed M. Ahmed and Salman Azhar, Miami, Florida, USA, April 25-26, 2002, pp.295-302.

Aoieong Raymond T., Tang, S.L. and Ahmed, Syed, M. (2003). "Process Cost Model for Construction Quality Measurement: Feedback and Case Study". *Proceedings of the Second International Conference on Construction in the 21st Century (CITC-II) on Sustainability and Innovation in Management and Technology*, Hong Kong, 10-12 December, 2003, pp.56-61.

BSI (British Standards Institution). (1992). BS 6143: Part 1 *Guide to the economics of quality Part 1. Process cost model*. British Standards Publishing Limited, London, UK.

Gibson, P.R., Hoang, K. And Teoh, S.K. (1991). "An investigation into quality costs". *Quality Forum*, Vol. 17, No. 1, pp.29-39.

Tang, S.L., Aoieong, T. Raymond and Ahmed, Syed M. (2004). "The use of process cost model (PCM) for measuring quality costs of construction projects: model testing". *Construction Management and Economics*, Vol.22, No. 3, pp.263-275.

Winchell, W.O. and Bolton, C.J. (1987). "Quality cost analysis: extend the benefits". *Quality Progress*, September 1987, pp.71-73.

11
APPLICATIONS OF THE CONSTRUCTION PROCESS COST MODEL

11.1 The Construction Process Cost Model Applied to Concreting Process

The construction process cost model (CPCM) is the name given for the PCM when it is applied to construction processes. As mentioned in Chapter 10, it is not used for capturing quality costs of an entire construction project but for capturing that of a particular process. This is in line with the "process approach" and "continual improvement" concepts of the latest (year 2000) version of the ISO 9000 quality management system, which is a step closer to the concept of Total Quality Management (Aoieong and Tang, 2002). The "continual improvement" element of CPCM will be illustrated in the later case studies. In applying CPCM, a construction process must be identified. In theory, CPCM can be applied to any construction processes, but the "concreting process" has been chosen for illustration in this chapter because it is the most common and well known process in construction projects. Figure 11.1 — a reproduction of Figure 10.2 — shows how CPCM is applied to the "concreting process". The boundary defined for the process includes formwork placing, reinforcement placing and concrete placing.

Figure 11.1 Process model for concreting

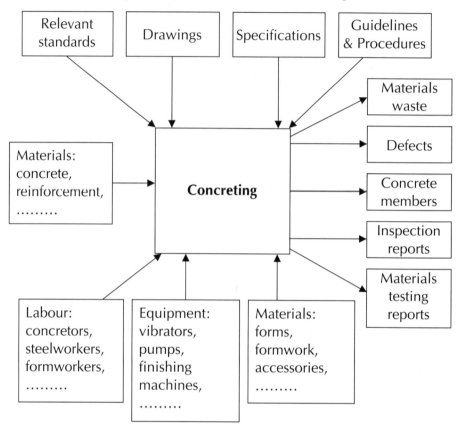

The quality costs in CPCM are called process costs, which can be divided into two parts: the costs of conformance (COC) and the costs of non-conformance (CONC) (see Chapter 10). COC is the intrinsic costs involved for providing the finished concrete product as required in good order, and the CONC is the costs of wasted time, materials and resources and any costs associated with the rectification of the unsatisfactory concrete product. In applying CPCM to capture the quality costs (or the process costs) of the concreting process, data on COC and CONC have to be collected from construction sites.

There are two basic and important principles for CPCM application: (1) CPCM must not be complicated and be easy to understand and use by site personnel of ordinary educational or technical level, and (2) CPCM must function as a tool for continual improvement of the process. These will be fully discussed in the two case studies presented in the following sections.

11.2 Case Study 1: A Building Project in Hong Kong

This project is the construction of two 38-storey high residential blocks housing with 760 residential units. The gross area for each typical floor is 591 square metres. The contractor is a reputable, ISO 9000: 2000 certified, construction firm, founded about 40 years ago in Hong Kong, with an annual turnover of about US$1 billion. The data collection started in the later part of 2001, which included COC and CONC for 18 typical floors (floors 21 through 38). The quantity surveyor of this project was responsible for providing the COC data because he could retrieve them from the contract documents, and the site engineer was responsible for providing the CONC data because he could conveniently record down the defects during his routine inspections. The COC provided by the quantity surveyor for each typical floor is shown in Table 11.1. He also indicated that it was difficult to separate the costs of labour and material from the information available in the contract document, because lump sum contracts were signed between the main contractor and subcontractors. The most significant cost item under equipment was the tower crane, which had been singled out in the cost items. Other equipment items (e.g. vibrators, compressors, etc.) were not known and were assumed to have been absorbed in the quoted prices. The ignorance of the costs of other equipment items is not harmful at all to the application of CPCM. After all, any non-conformance costs related to defects caused by the inadequacy or misuse of equipment will be counted as CONC.

Table 11.1 Cost of conformance (COC) per typical floor

	Cost (HK$)	Remark
Formwork	321,505	Labour + Materials
Reinforcement	103,537	Labour + Materials
Concrete	162,290	Labour + Materials
Equipment	10,282	Tower crane
Material testing	225	Reinforcement
	1,600	Concrete
Total Cost of conformance: 599,439		($1,014/sq. m.)

The CONC is not as easy to acquire as the COC. The acquisition of CONC is to be described in three parts: formwork placing, reinforcement placing, and concrete placing.

11.2.1 Formwork Placing

For formwork placing, the accuracy of the locations and dimensions of concrete members depends on the proper setting out and the proper support of the formwork. Major errors in setting out the location of formwork may result in a complete knock down and relocation of the concrete members. Fortunately, such problems usually are not expected to happen in typical floor construction because of its repetitive nature and of the close supervision by different parties. On the other hand, minor errors in setting out and slight formwork movement may happen quite often and it may lead to significant repair costs. For example, a concrete wall with a few millimetres off the centreline may result in extensive chiselling and cleaning to the finished concrete work. Proper written records of such defects are usually done by foremen and/or site engineers and kept in the site office for reference. As observed by the authors, typical defects were usually caused by insufficient supports of formwork that led to slight movement at the bottom during concrete placing. The related CONC can be calculated based on the estimated time and labour required for fixing each non-conformance occurrence. Data collected for CONC are shown in Table 11.2.

Table 11.2 Cost of concrete repair works caused by formwork movement

Floor	No. of occurrence	Time to repair (hrs.) *	Cost to repair (HK$) **
21	10	20	1,800
22	13	26	2,340
23	8	16	1,440
24	8	16	1,440
25	11	22	1,980
26	12	24	2,160
27	10	20	1,800
28	4	8	720
29	3	6	540
30	9	18	1,620
31	7	14	1,260
32	9	18	1,620
33	10	20	1,800
34	8	16	1,440
35	11	22	1,980
36	12	24	2,160
37	9	18	1,620
38	7	14	1,260
Avg.			1,609

 *   Based on an estimated time for each occurrence of 2 hours (information provided by the site engineer).

**   Based on an hourly wage rate of HK$90 (prevailing wage of carpenters in Hong Kong).

11.2.2 Reinforcement Placing

For reinforcement placing in this case study, the cost of non-conformance for the reinforcement placing process included only those costs incurred due to defects found during the final inspection before concrete pouring. In order to compile a list of common defects found in reinforcement placing, the author had spent extensive time in observing the inspection process and discussing with the resident engineer. (It is a regulation in Hong Kong to have a resident engineer, a representative from the consulting firm, stationed on site to supervise a contractor's work.) It was concluded that non-conformances were mainly caused by poor workmanship of the subcontractor. A form containing a checklist of all the common defects was then designed to facilitate the site staff in the data collection process. A typical form used to collect the defect data for each floor is shown in Table 11.3. The number of occurrence of each type of defects was recorded and the time and cost required for the remedial work were then estimated based on current labour rate. These are also shown in Table 11.3. The CONC data for floors 21 through 38 are shown in Table 11.4.

11.2.3 Concrete Placing

For concrete placing, the site engineer indicated that honeycombing caused by insufficient compaction of freshly placed concrete was the most common non-conformance that required extensive repair work. After the formwork was struck, the resident engineer representing the consulting firm would inspect all concrete surfaces. Based on the judgment of the resident engineer, concrete honeycombs that required remedial works would be marked and photographed. The honeycombs would then be categorized into Type 1 and Type 2 according to the severity of the defects as shown in Table 11.5. The time and cost required to complete the remedial work of each type of honeycombs were then estimated based on the current labour and material rates as shown in Table 11.6.

The CONC of each typical floor for all the three processes is summarized in Table 11.7. A typical process cost report for the 31st floor is also shown in Table 11.8 (see Table 10.2 of Chapter 10). The total process cost (COC + CONC) of each typical floor from 21/F to 38/F is tabulated and represented in Table 11.9. Due to certain missing data from the site, process cost for the 30th floor was not recorded.

Table 11.3 CONC of reinforcement placing for a typical floor

Number of occurrences	30/F-31/F wall	31/F slab/ beam	Time (mins.)	Cost * (HK$)
Date of inspection				
WALL Links				
links missing				
links not properly tied				
links not properly spaced				
cover not enough	5		5 × 10	100
WALL Main Bars				
wrong size				
bars missing				
bars not properly spaced	5		5 × 15	150
bar length not enough				
lap length not enough				
BEAM Stirrups				
stirrups missing		2	2 × 15	60
stirrups not properly tied		17	17 × 5	170
stirrups not properly spaced		10	10 × 10	200
cover not enough		5	5 × 10	100
BEAM Main Bars				
wrong size				
bars missing				
bars not properly spaced		5	5 × 15	150
bar length not enough				
lap length not enough				
SLAB				
bars in AC hood not properly located		2	2 × 10	40
bars in fins not properly located				
bars in wrong layer				
bars in wrong direction				
bars not properly spaced		6	6 × 15	180
cover not enough		4	4 × 10	80
chairs not properly located		3	3 × 5	30
OTHERS				
construction joints N.G.		1	1 × 15	30
starter bars not cleaned		5	5 × 10	100
clean up works required		5	5 × 10	100
Total cost of non-conformance:				**$1,490**

* Based on an hourly wage rate of HK$120 (prevailing wage for steel workers)

Table 11.4 CONC of reinforcement placing for floors 21 through 38

Floor	CONC (HK$)
21/F	1,060
22/F	1,320
23/F	1,420
24/F	1,120
25/F	1,250
26/F	960
27/F	1,250
28/F	1,640
29/F	1,750
31/F	1,490
32/F	1,260
33/F	1,370
34/F	1,480
35/F	930
36/F	1,160
37/F	1,180
38/F	1,170
Avg.	**$1,283**

Table 11.5 Concrete honeycomb categorization

Defect type	Estimated time to complete (1 labour per occurrence)		Total time
	Chiselling/cleaning	Repair	
Type 1: Without formwork involved, e.g. typical honeycombs at faces/bottom of wall and at corners of window openings	10 mins.	5 mins.	0.25 hour
Type 2: With formwork involved, e.g. typical honeycombs at top of wall/column with reinforcement totally exposed	30 mins.	120 mins.	2.50 hours

Table 11.6 Cost estimation of concrete honeycombs for typical floors

Floor	Type 1			Type 2			Total cost to repair (HK$)
	No.*	Total time to repair (hrs.)	Total cost to repair** (HK$)	No.*	Total time to repair (hrs.)	Total cost to repair** (HK$)	
21	1	0.25	23	0	0	0	23
22	2	0.50	45	1	2.50	225	270
23	3	0.75	68	0	0	0	68
24	3	0.75	68	0	0	0	68
25	2	0.50	45	1	2.50	225	270
26	1	0.25	23	0	0	0	23
27	4	1.00	90	1	2.50	225	315
28	4	1.00	90	1	2.50	225	315
29	2	0.50	45	0	0	0	45
30	6	1.50	135	0	0	0	135
31	4	1.00	90	1	2.50	225	315
32	2	0.50	45	0	0	0	45
33	5	1.25	113	2	5.00	450	563
34	6	1.50	135	1	2.50	225	360
35	4	1.00	90	0	0	0	90
36	8	2.00	180	1	2.50	225	405
37	7	1.75	158	1	2.50	225	383
38	6	1.50	135	0	0	0	135
Avg.							$217

*   Number of occurrence observed after the inspection.
** Based on an hourly wage rate of HK$90.

Table 11.7 Cost of non-conformance of each process

Floor	Cost of Non-conformance (HK$)						Total (HK$)
	Formwork placing		Reinforcement placing		Concrete placing		
21/F	1,800	62%	1,060	37%	23	0.8%	2,883
22/F	2,340	60%	1,320	34%	270	6.9%	3,930
23/F	1,440	49%	1,420	48%	68	2.3%	2,928
24/F	1,440	55%	1,120	43%	68	2.6%	2,628
25/F	1,980	57%	1,250	36%	270	7.7%	3,500
26/F	2,160	69%	960	31%	23	0.7%	3,143
27/F	1,800	53%	1,250	37%	315	9.4%	3,365
28/F	720	27%	1,640	61%	315	11.8%	2,675
29/F	540	23%	1,750	75%	45	1.9%	2,335
31/F	1,260	41%	1,490	49%	315	10.3%	3,065
32/F	1,620	55%	1,260	43%	45	1.5%	2,925
33/F	1,800	48%	1,370	37%	563	15.1%	3,733
34/F	1,440	44%	1,480	45%	360	11.0%	3,280
35/F	1,980	66%	930	31%	90	3.0%	3,000
36/F	2,160	58%	1,160	31%	405	10.9%	3,725
37/F	1,620	51%	1,180	37%	383	12.0%	3,183
38/F	1,260	49%	1,170	46%	135	5.3%	2,565
Avg.	$1,609	51%	$1,283	42%	$217	7%	$3,110

Table 11.8 Process cost report for the concreting process of the 31st floor

Process cost report Process name: Concreting Boundary: "Formwork construction" to "Formwork striking" 31th floor of Building XXX Process owner: various							
Process Conformance Activity:	Cost (HK$) Act.	Est.	Process Owner	Process Non-conformance Activity:	Cost (HK$) Act.	Est.	Process Owner
(Formwork erection) • Labour & materials	321505		Formwork subcontractor	(Formwork erection) • concrete rework due to formwork movement	1260		Formwork subcontractor
Activity: (Reinforcement) • Labour & materials	103537		Steel subcontractor	Activity: (Reinforcement) • rework due to non-conformance	1490		Steel subcontractor
Activity: (Concrete placing) • Labour & materials • testing	162290		Concrete subcontractor	Activity: (Concrete placing) • rework due to honeycombs in concrete	315		Concrete subcontractor
Others: • Material testing • Equipment	1825 10282		General contractor				
Total process conformance cost	$599,439			Total process non-conformance cost	$3,065		

Table 11.9 Total process cost for typical floors

Floor	Cost of Conformance (HK$) *		Cost of Non-conformance (HK$) **		Total Process Cost (HK$)
21/F	599,439	99.52%	2,883	0.48%	602,322
22/F	599,439	99.35%	3,930	0.65%	603,369
23/F	599,439	99.51%	2,928	0.49%	602,367
24/F	599,439	99.56%	2,628	0.44%	602,067
25/F	599,439	99.42%	3,500	0.58%	602,939
26/F	599,439	99.48%	3,143	0.52%	602,582
27/F	599,439	99.44%	3,365	0.56%	602,804
28/F	599,439	99.56%	2,675	0.44%	602,114
29/F	599,439	99.61%	2,335	0.39%	601,774
31/F	599,439	99.49%	3,065	0.51%	602,504
32/F	599,439	99.51%	2,925	0.49%	602,364
33/F	599,439	99.38%	3,733	0.62%	603,172
34/F	599,439	99.46%	3,280	0.54%	602,719
35/F	599,439	99.50%	3,000	0.50%	602,439
36/F	599,439	99.38%	3,725	0.62%	603,164
37/F	599,439	99.47%	3,183	0.53%	602,622
38/F	599,439	99.57%	2,565	0.43%	602,004
Avg.	$599,439	99.48%	$3,110	0.52%	$602,549

*   Obtained from Table 11.1
** Obtained from Table 11.7

Figure 11.2 shows the plotting of points for each process cost against each floor (or cycle) using the figures from Table 11.9. The slope of the linear regression line of these points is found to be negative. (The regression equation is: $y = -0.272x + 602557$). This means that the process cost of each floor is decreasing when the level of the floor (or cycle) is going up. This is a good phenomenon because the concreting process is continually improving. The process cost therefore can serve as a tool for the indication of continual improvement. By referencing the process costs, the contractor may know whether or not his own effort put in the work is sufficient and, hence, may develop appropriate strategies for achieving continual improvement.

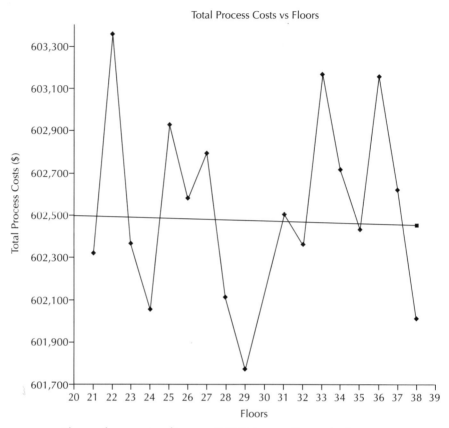

Slope of regression line = − 0.272 (a negative value)

Figure 11.2 Linear regression of total process costs vs floors

11.3 Discussions for Case Study 1

11.3.1 Completeness of Non-conformance Data

Due to the insufficiency of human resources for recording non-conformances in the "reinforcement placing" process, only those discovered during the final inspection by the resident engineer on each floor were recorded. However, defects might be discovered by the contractor's foremen or subcontractors' foremen at any time during the process, and these were not recorded in this work. In order to obtain a full picture of all the defects occurring in the "reinforcement placing" process, their involvement in the data collection process is essential. So, the CONC recorded in "reinforcement placing" in the case study may be slightly underestimated. The non-conformances occurring in the concrete placing process were mainly honeycombing due to insufficient compaction of concrete. Since all related records for concreting were properly kept, the cost of non-conformance could be accurately estimated. Similarly, all records related to the remedial concrete works caused by improper alignment of formwork were also kept in the site office and, therefore, the related CONC can be easily retrieved and accurately estimated.

11.3.2 Involvement of the Resident Engineer

As mentioned earlier, it is a regulation in Hong Kong to have a resident engineer, a representative from the consultant firm, stationed on site to supervise the contractor's work. It is also a common practice in Hong Kong for the resident engineer to inspect and ensure that all reinforcements are in conformance to the structural drawings prior to concrete placing. Since the resident engineer's inspection is fairly thorough and impartial, his or her involvement in the data collection process will definitely enhance the quality of data recorded. However, though it is essential to get the involvement of all parties: the resident engineer (consulting firm), the site engineer (contractor) and the subcontractors to collect a complete set of data, the authors observed that these three parties were quite often in conflict among each other due to the inherent differences in their emphases, i.e. time, cost and quality. While the contractor's main concern is to get the work done as quickly as possible (time), the resident engineer, however, would like to see that every piece of work is done according to the standards and specifications (quality). On the other hand, the subcontractor's main concern is to make the highest profit (cost). Therefore, care must be taken to reduce conflicts among the three parties before they get involved in the data collection process. The recent concept of 'partnering' may be able to help reduce such conflicts.

11.3.3 Short Cycle Time

The cycle time between the constructions of each typical floor may have a significant impact on the site staff's attitude towards quality. The targeted cycle time for this project was four days. This is something very peculiar and can only, as far as the authors know, happen in Hong Kong. Since the project was under such a tight schedule, the site staff, including the foremen and the engineers, had to work considerable overtime in order to meet the target date. During casual conversation with the site staff, the authors had observed that most workers would consider quality only when they were not under a tight schedule. In other words, allowing reasonable time for workers to finish their work is a prerequisite for the success of any quality initiatives at the site level. The authors agree with this view of the workers.

11.3.4 Process Cost for the Construction of Typical Floors

Based on the data collected, the average process cost of the concreting of a typical floor was $602,549. When it was expressed in terms of unit floor area, the process cost was $1020/sq.m. The cost of non-conformance ranged from 0.39% to 0.65% of the total process cost. It can be seen that the cost of non-conformance for this particular process is quite insignificant for the following reasons:
- Both the concretors and steel fixers of this project were highly skilled workers.
- For reinforcement placing, only non-conformances observed in the final inspection were recorded. The CONC therefore were underestimated.
- The use of aluminum formwork greatly reduced defects caused by traditional timber formwork.

An average of 51% of the cost of non-conformance was related to improper placing and supporting of formwork. As a result, the labour cost in chiseling finished concrete was extensive.

11.3.5 Continual Improvement

The continual improvement observed in the case study could possibly be a result of the fact that the contractor and the subcontractors knew at the beginning that this project was to be used for CPCM testing. Therefore, it is possible that for a project when the contractor knows that his work is going to be monitored by CPCM, he will probably be more careful and pay more attention to the quality of his work.

11.4 Case Study 2: A Design Project

The CPCM model is a method for applying the concept of quality costing to any process or service. Since construction projects are closely related to design processes, a design project was selected as a case study in an attempt to apply the concept of quality costing to processes other than construction processes. A typical CPCM for design processes is shown in Figure 11.3. The project is an entertainment and theme park complex development in

China. The total cost of this 100,000 square metre complex is estimated to be US$110 million. The design and construction of this project is managed by a well-known international contractor.

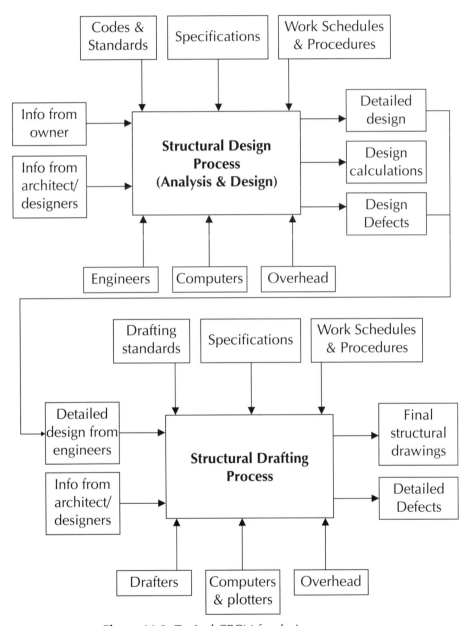

Figure 11.3 Typical CPCM for design processes

The major cost item relating to the design processes is salary. Specific resources utilized by engineers and draftsmen in design offices are software, computers and plotters. Typical data required for the estimation of the cost of conformance (COC) of a particular day are the daily man-hours allocated to the project, computer usage time and the number drawings plotted. Both the engineers and the draftsmen involved in the project were requested to fill up a time sheet with the above data. With the Director's input of the staff's wages and equipment costs, the daily COC of the design process was calculated. The COC of a particular day is shown in Table 11.10 and the total COC of up to the end of the data collection period is summarized in Table 11.11.

Table 11.10 COC on a particular day (7 June 2003)

	Man-hour/ (hour)	Number (drawing)	Rate ($/hr. or $/drwg.)	Cost
Design	4		100	400
Drafting	4		55	220
Computer usage	3		35	105
Plotting		3	20	60
Total cost of conformance:				785

Table 11.11 COC of the design project

Date	Design ($)	Drafting ($)	Computer Usage ($)	Plotting ($)	COC ($)
May 30	200	0	70	0	270
May 31	400	0	140	0	540
:	:	:	:	:	:
June 7	400	220	105	60	785
:	:	:	:	:	:
Aug. 8	800	550	263	0	1613
Aug. 9	800	220	105	0	1125
Aug. 11	800	220	140	40	1200
Total	21,600	7,150	4,323	160	33,233

The discussion of cost of non-conformance (CONC) is divided into two parts: design and drafting.

11.4.1 Design

Structural design is a process which converts the conceptual design of architects into buildable structural products. The design process usually involves inputs from the owner and the architect of the project as shown in Figure 11.3. Typical outputs are sketches, details and calculations and these in turn become the inputs to the drafting process. Common causes of defect in the design process are wrong inputs from architects and errors made by engineers in the computing process. Reworks due to changes originating from owners/architects and faults from constructors are also very common in design offices. Strictly speaking, the cost of reworks due to changes originating from owners, architects or constructors should not be included in the CONC calculation because these changes are not due to non-conformances. However, when claims for extra costs due to changes from the owner are not allowed due to contractual agreements between the owner and the designers, the cost has to be absorbed by the designers. Under this circumstance, both the related COC and CONC had to be included in the calculation of process cost. Since the Director of the design office indicated that no claims for extra costs due to changes were to be allowed, it was decided to include cost of reworks due to changes in the COC and CONC calculations. In order to minimize the extra time required for the design team for reporting, a list of causes of defect and rework was incorporated in the time sheet commonly used in design offices. Once the times for reworks were known, the CONC could be estimated based on the staff's wages. An additional computer usage cost, based on the estimated time provided by the designer, was also included in the CONC calculation. A typical form for the collection of data and the CONC is shown in Table 11.12. The total CONC of the design process for the period studied is shown in Table 11.13.

11.4.2 Drafting

Drafting is a process which converts the inputs from engineers and architects into a set of working drawings submitted to the owner as a partial fulfillment of the contractual requirements. Similar to the design process, a list of common causes of rework during the drafting process can be incorporated in the draftsmen's time sheet and the CONC can be estimated. These are shown in Table 11.14. The total CONC of the drafting process for the period studied is shown in Table 11.15.

Summaries of the daily and total process costs (COC + CONC) captured during the period studied are presented in Table 11.16 and Table 11.17 respectively. The average daily CONC, expressed as a percentage of the daily process cost, is found to be 16%. A typical process cost report for the period studied is shown in Table 11.18. The daily COC, CONC and the total process costs (COC + CONC) captured within the period studied are shown in Figure 11.4.

Table 11.12 CONC of the design process

Date: June 6, 2003

Reasons for reworks	Time for reworks (hr.)	Cost * ($)
computational errors	4	400
incorrect detailing	—	
incorrect methodologies	—	
incorrect input to computer	—	
incorrect input from architects	—	
changes from architects	—	
others (please specified)	—	
Sub-total:	4	400
Computer usage**	1	35
Total:		435

* Based on an hourly wage rate of $100.
** Based on 25% of the time for reworks.

Table 11.13 Total CONC of the design process

Date: May 30 to August 11, 2003

Reasons for reworks	Time for reworks (hr.)	Cost * ($)
computational errors	10	1000
incorrect detailing	0	0
incorrect methodologies	0	0
incorrect input to computer	16	1,600
incorrect input from architects	12	1200
changes from architects	9	900
others (please specified)	1	100
Sub-total:	4	400
Computer usage**	12	420
Total:		5220

* Based on an hourly wage rate of $100.
** Based on 25% of the time for reworks.

Table 11.14 CONC of the drafting process

Reasons for reworks	Time for reworks (hr.)	Cost * ($)
Drafting (operational) errors	—	
incorrect detailing	—	
incorrect methodologies	—	
incorrect input to computer	—	
incorrect input from architects	4	
changes from architects	—	220
others (please specified)	4	
Sub-total:	3	220
Computer usage**		105
Total:		325

* Based on an hourly wage rate of $55.
** Based on 75% of the time for reworks.

Table 11.15 Total CONC of the drafting process

Date: June 10 to August 11, 2003

Reasons for reworks	Time for reworks (hr.)	Cost * ($)
Drafting (operational) errors	0	0
incorrect input to computer	0	0
incorrect input from designers	4	220
incorrect input from architects	0	0
changes from architects	34	1,870
others (please specified)	0	0
Sub-total:	38	2,090
Computer usage**		998
Total:		3,088

* Based on an hourly wage rate of $55.
** Based on 75% of the time for reworks.

Table 11.16 Summary of daily COC, CONC, and Process Cost (COC+CONC)

Date	Daily Cost of Conformance		Daily Cost of Non-Conformance				Daily Process Cost
	COC	% of total	Design	Drafting	Sub-total	% of total	
30-May	270	55.4%	217.5	0	217.5	44.6%	487.5
31-May	540	71.3%	217.5	0	217.5	28.7%	757.5
2-Jun	135	100.0%	0	0	0	0.0%	135
3-Jun	1080	71.3%	435	0	435	28.7%	1515
4-Jun	1080	83.2%	217.5	0	217.5	16.8%	1297.5
5-Jun	940	68.4%	435	0	435	31.6%	1375
6-Jun	925	68.0%	435	0	435	32.0%	1360
7-Jun	785	100.0%	0	0	0	0.0%	785
8-Jun	650	100.0%	0	0	0	0.0%	650
9-Jun	1190	100.0%	0	0	0	0.0%	1190
10-Jun	925	74.0%	0	325	325	26.0%	1250
11-Jun	725	69.0%	0	325	325	31.0%	1050
12-Jun	900	89.2%	108.75	0	108.75	10.8%	1008.75
13-Jun	800	100.0%	0	0	0	0.0%	800
14-Jun	400	100.0%	0	0	0	0.0%	400
16-Jun	1450	100.0%	0	0	0	0.0%	1450
17-Jun	1450	76.9%	435	0	435	23.1%	1885
18-Jun	1450	62.5%	870	0	870	37.5%	2320
19-Jun	725	62.5%	435	0	435	37.5%	1160
20-Jun	710	100.0%	0	0	0	0.0%	710
21-Jun	325	50.0%	0	325	325	50.0%	650
23-Jun	440	100.0%	0	0	0	0.0%	440
25-Jul	800	88.0%	108.75	0	108.75	12.0%	908.75
26-Jul	400	100.0%	0	0	0	0.0%	400
28-Jul	800	100.0%	0	0	0	0.0%	800
29-Jul	800	100.0%	0	0	0	0.0%	800
30-Jul	800	100.0%	0	0	0	0.0%	800
31-Jul	800	100.0%	0	0	0	0.0%	800
1-Aug	800	100.0%	0	0	0	0.0%	800
2-Aug	400	100.0%	0	0	0	0.0%	400
4-Aug	1450	100.0%	0	0	0	0.0%	1450
5-Aug	1450	100.0%	0	0	0	0.0%	1450
6-Aug	1450	100.0%	0	0	0	0.0%	1450
7-Aug	1450	69.0%	0	650	650	31.0%	2100
8-Aug	1612.5	56.4%	435	812.5	1247.5	43.6%	2860
9-Aug	1125	59.7%	435	325	760	40.3%	1885
11-Aug	1200	61.2%	435	325	760	38.8%	1960
Average	898	84%	141	83	224	16%	1123

Table 11.17 Total process cost during the period studied

Cost of Conformance		Cost of Non-conformance		Total CONC		Total Process Cost
		Design	Drafting			
33,233	80%	5,220	3,088	8,308	20%	41,541

Applications of CPCM (Construction Process Cost Model)

Table 11.18 Process cost report for the design process during the period studied

Process cost report Process name: Design Boundary: "Designing" to "Drafting" Period studied: May 30 to August 11, 2003 Process owner: Consulting company							
Process Conformance	Cost Act.	Est.	Process Owner	Process Non-conformance	Cost Act.	Est.	Process Owner
Activity: Design • Labour	21600		Consulting Company	Activity: Design • computational errors • incorrect input to computer • incorrect input from architects • changes from architects • constructor's faults	1000 1600 1200 900 100		Consulting Company
Activity: Drafting • Labour	7150		Consulting Company	Activity: Drafting • incorrect input from designers • changes from architects	220 1870		Consulting Company
Others: • Computer • Plotter		4323 160	Consulting Company	Others: • Computer		1418	Consulting Company
Total process conformance cost	$33,233			Total process non-conformance cost	$8,308		

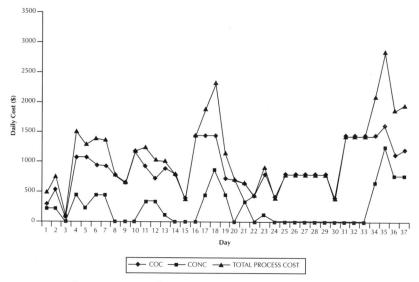

Figure 11.4 Daily COC, CONC and total process cost

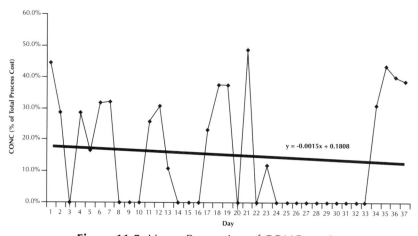

Figure 11.5 Linear Regression of CONC vs. time

Figure 11.5 shows the plotting of points for the daily CONC expressed as a percentage of the daily total process cost using the data from Figure 11.4. The slope of the linear regression line of these points is found to be negative. (The regression equation is: $y = -0.0015x + 0.1808$ and the line intercepts y-axis at 18%.) This means that the daily CONC, when expressed as a percentage of the daily total process cost, is decreasing with time. This is a good phenomenon because there is an improvement with time. The process cost therefore can serve as a tool for the indication of continual improvement. By referencing the process costs, the consultant may know whether or not his or her own effort put into the work is sufficient and, hence, may develop appropriate strategies for achieving continual improvement.

11.5 Discussions for Case Study 2

11.5.1 Completeness of Non-conformance Data

Unlike the construction processes where the subcontractors involved might be numerous, the design process is usually performed under one roof. Therefore, the number of personnel involved in monitoring the design process is much reduced. Since the working environment in a design office is far more pleasant than that at a construction site, personnel in design offices in general are usually more cooperative and willing to participate in case studies. As a result, the non-conformance data obtained in this study was considered more complete and accurate. Nevertheless, a full support from the top management is still a prerequisite for the successful implementation of CPCM.

11.5.2 Phased Construction

The delivery method of this project takes the form of a construction management (CM) contract in which an internationally contractor has a contract with the developers to project manage the design and construction of the project. One advantage of this type of contract is time saving in the overall project duration because of the overlapping of the design and construction phases (Tang *et al.* 2003). However, the schedule of this type of fast-track construction must be well-controlled or else serious delays in design and/or construction may occur. Due to the tight construction schedule of the project under study, the construction of the foundation had to commence first, even before the architectural plans had been finalized and issued. As a result, the foundations were designed without the complete details of the superstructure from the architect. During the monitoring period of the case study, the design process was put on hold and interrupted for a period of four weeks. One of the reasons for this interruption was the fact that the architect could not catch up with the schedule. The process cost of the design process during this period of time was recorded as nil because the design effort was shifted to other projects. As a consequence, the COC would be enormous if other projects could not have been assigned to the design team during this interrupted period.

11.5.3 Owner's or Architect's Changes and Project Manager's Coordination

As discussed above, the tight construction schedule resulted in a delay in the architectural design. Another cause of delay was the incomplete concept from the owner. The latter is always a risk when an owner has to accept construction works commencing before the conceptual design has been finalized. If the rework costs incurred by the architect/engineers due to the changes made by the owner can be claimed, the original design process cost will remain unchanged. However, the design process cost will be increased by the amount of rework due to the changes if no claim is allowed.

11.5.4 Constructor's Faults

Similar to the changes made by the owner during the design process, the constructor's faults may also result in reworks in the design process. Since changes due to constructor's faults are inevitable in any construction project, it is normal practice for design offices to absorb the cost of reworks provided that it is not too significant. The implementation of CPCM in design offices enables the tracking of this type of rework so that proper actions can be taken before the problems become out of control.

11.5.5 Process Cost for the Design Process

Based on the available data, the total process cost for the structural design and drafting was $41,541. The cost of non-conformance was 20% of the total process cost (see Table 11.17). The reason for this high percentage of CONC was that numerous changes were made by the architect of the project after the design and drafting were completed. As a result, a substantial part of the design and drafting had to be done again. As mentioned previously, the costs of reworks due to changes originating from the owners, architects or constructors should not be included in the CONC calculation because these changes are not due to non-conformance. In this case study, the items "changes from architect" and "incorrect input from architect" amounted to 62% of the total CONC. Although claims for extra costs due to changes from the owner are not allowed in this consultancy agreement, the top management of the consulting company should be alerted because this amount has to be absorbed by the company. If the items "change from architect" and "incorrect input from architect" were not categorized as reworks, the CONC would be drastically reduced to 8.6% of the total process cost. This figure reflected the amount of rework originating from the consultant's own non-conformance alone.

11.6 Other Possible Applications of CPCM

Two cases, one on concreting process and another on design process, have been described in detail in this chapter. As a matter of fact, CPCM can be applied to many other processes in the construction industry. A study of the application of CPCM to other construction related processes as well as management processes is recommended. Management processes such as project planning and monitoring, procurement, resource allocation, staff training and site safety are also worthy of investigation. Actually, the application of CPCM can be extended to services such as inspection, testing and installation. The versatility and practicability of CPCM makes it a promising and essential tool in measuring improvement in the construction industry.

References

Aoieong Raymond T. and Tang, S.L. (2002). "The Year 2000 Version of ISO 9000 and the Process Cost Model for measuring Quality Costs in Construction Processes", *Proceedings of the first International Conference on Construction in the 21st Century*, Editors: Irtishad Ahmad, Syed M. Ahmed and Salman Azhar, Miami, Florida, USA, April 25-26, 2002, pp.295-302.

Aoieong, Raymond T., Tang, S.L. and Ahmed, Syed M. (2002). "A process approach in measuring quality costs of construction projects: model development". *Construction Management and Economics*, Vol. 20, No. 2, pp.179-192.

Aoieong, Raymond T., Tang, S.L. and Ahmed, Syed M. (in press). "Construction process cost model (CPCM) applied to construction and design process for capturing quality costs". Paper being reviewed by *ASCE Journal of Construction Engineering and Management for publication*.

Tang, S.L., Aoieong, Raymond T. and Ahmed, Syed M. (2004). "The use of process cost model (PCM) for measuring quality costs of construction projects: model testing". *Construction Management and Economics*, Vol. 22, No.3, pp.263-275.

Tang, S.L., Poon, S.W., Ahmed, Syed M. and Wong, K.W. Francis (2003). *Modern construction project management*. Hong Kong University Press, Hong Kong. Chapter 3.

12
DEVELOPING A QUALITY CULTURE AS THE WAY FORWARD

One of the main emphases of the quality management system is continual improvement of an organization, in particular, the effectiveness of its processes and its products. In order to identify areas for improvement, quantitative measurements are necessary. The tool, quality costing — particularly the construction process cost model (CPCM) discussed in Chapters 10 and 11 — is considered as one of the quality improvement techniques which can be applied to the construction industry. However, having a correct tool to control and measure continual quality improvement is only the first step towards a successful implementation of a quality management programme. There are many key factors without which a quality management system will fail. Among them, continual commitment from top management and employees is described by many researchers as the most important factor (Burati Jr. and Oswald 1993; Carlsson and Carlsson 1996; Low and Omar 1997; Mo and Chan 1997; Tan 1997; Arditi and Gunaydin 1998; Laszlo 1999, Ahmed *et al.* 2005). This can be understood because though an organization has the best quality plan prepared by consultants, the success still depends on how the organization (top management and employees) implements it. The collective attitudes and beliefs of employees towards quality are commonly described as the quality culture of an organization. Therefore, the quality management system and its principles must build on a good quality culture.

12.1 Quality Culture

Quality culture is defined as the pattern of human habits, beliefs, and behavior concerning quality (Gryna 2001). An organization with a good "quality culture" is the one having positive and clear habits, beliefs, and behavior concerning quality. These habits, beliefs, and behavior will manifest themselves in the actions of top management and employees. Therefore, creating a good quality culture that supports quality management is vital in the implementation and maintenance of quality management systems. Otherwise, as quoted from Cortada and Woods (1995), "the use of quality management techniques will be mechanical and not likely to deliver the long-term results for customer satisfaction, more efficient processes, lower costs, growth, and profitability that it promises".

Before developing a good quality culture in the construction industry, a brief review of the existing quality culture is necessary. Due to the complexity of construction projects, the development of the existing quality culture in the construction industry can be described as the result of certain inherent characteristics in construction projects. Ritz (1994) describes four common characteristics of construction projects as follows:

1. Each project is unique and not repetitious.
2. A project works against schedules and budgets to produce a specific result.
3. The construction team cuts across many organizational and functional lines that involve virtually every department in the company.
4. Projects come in various shapes, sizes, and complexities.

Each characteristic above shapes the quality culture of the construction industry to a certain extent. As each project is unique to the parties involved, most people will not be staying in one project for too long a period. This high mobilization of construction staff will result in shortsightedness and a lack of long-term commitment of staff towards quality in general. Moreover, the bonding between contractor and subcontractors and among subcontractors is usually weak due to the relatively short duration of construction projects. Therefore, it is rather difficult to create a teamwork environment on site. Construction projects usually have a prescribed scope, schedule and budget to produce a quality "product". Balancing the triangular relation between the scope, schedule and budget of a project is always a challenge faced by management. Though quality is generally viewed as an integral part of scope, budget and schedule, its importance can easily be neglected due to the ever changing of scope from clients and tight budget and schedule resulted from the highly competitive construction market. Due to the complexity of construction projects, the parties involved, subcontractors in particular, in a particular project is usually numerous. In Hong Kong, it is not uncommon for a construction project to have more than 50 subcontractors on site, with more than 80% of work in terms of contract sum to be subcontracted (Tang et al. 2003). It is still debatable whether multi-layer subcontracting is beneficial or adverse to the construction industry. In the context of quality management, however, its adverse effects are obvious. First of all, it is natural for an individual party involved to put its own interest before others. As a result of this, there will be a lack of communication, trust and harmony among different parties and the atmosphere is usually confrontational and adversarial. Moreover, multi-layer subcontracting reduces the profit margin of each party and therefore leaving no extra resources for the implementation of any quality initiatives. Meeting the minimum quality requirement from customers becomes an accepted culture in the industry in general because of the financial constraint. The complexity of construction projects also shapes the quality culture of the construction industry. In a typical construction site, it is common to have many construction processes and activities going on simultaneously. It is essential to monitor and control these processes besides the final products because the cost of remedial work of nonconforming products is usually very substantial. Unfortunately, it is a common practice to inspect only the final product at different stages of the construction due to resource constraint.

Broadly speaking, the above discussion has pointed out that the "natural" quality culture existed in the construction industry has been in constant conflict with the key features of quality management systems advocated by researchers. Chase and Federle (1992), Abdul-Rahman (1996) and Tsiotras and Gotzamani (1996) also indicated that one of the main reasons for the failure of any quality management initiatives was the lack of long-term commitment to quality from the top management. Moreover, numerous surveys and publications (Low and Yeo 1997; Ahmed and Aoieong 1998; Moatazed-Keivani et al. 1999; Kumaraswamy and Dissanayaka 2000) have concluded that "fulfilling clients' requirements", and not "improving quality", was considered by most contractors as one of the main motivations for undertaking any quality management initiatives. As described by Kam and Tang (1998), the Hong Kong contractors were forced, by their major clients, Hong Kong Works Bureau and Hong Kong Housing Authority, to develop their quality management systems based on the ISO 9000 standard. The ISO certification so obtained was also considered as a "work permit" to bid government projects. As a result of it, the driving force, or commitment, from the top management to review, maintain and improve the quality management system will soon diminish once the certification process is completed. Needless to say, the lack of incentive to do better will definitely be reflected from the employees' daily works. Once again, the findings reinstate the fact that the successful implementation of a quality system lies on how organizations implement it and not the system itself.

12.2 Quality Culture Audit

In order to create an environment which will facilitate the implementation of quality management systems, changes in quality culture are necessary. As a matter of fact, cultural change, whether small or drastic, is a prerequisite to the successful implementation of any quality programme. Since different organizations have different cultural pattern, it is essential for the top management to conduct a cultural audit for various levels of management and work force so that the differences between the existing and the desired quality culture can be properly assessed.

The cultural audit can be conducted through carefully designed questionnaire surveys on quality and interviews with top management and employees. The content of such surveys and interviews should include the evaluation of the values and attitudes of employees towards quality. In designing the questionnaire, the following characteristics of a good quality culture can be used as a benchmark (Goetsch and Davis 2000):

1. Open, continual communication
2. Mutually supportive internal partnerships
3. Teamwork approach to problems and processes
4. Obsession with continual improvement
5. Broad-based employee involvement and empowerment
6. Sincere desire for customer input and feedback

The cultural audit is an important step towards cultural change particularly in the construction industry due to the fact that construction worker's perceptions of quality is generally low. An example of an assessment worksheet for collecting information is also provided by Goetsch and Davis (2000). With the results obtained from the cultural audit as baseline, top management can properly identify the areas for changes, access the resources required and apply specific measures to cultural change which will be beneficial to the organization.

12.3 A Change of Culture

The successful implementation of quality management systems requires the creation of a quality culture. According to Cortada and Woods (1995), the five values below must be presented by the top management to create a quality culture:

1. A focus on customer satisfaction
2. A focus on processes and their continual improvement
3. A focus on teamwork and cooperation
4. A focus on openness and sharing of information
5. A focus on the use of scientifically derived data for making decisions

Similar views were shared by Laszlo (1999) and Ngowi (2000). As a matter of fact, these values are very much similar to the quality management principles on which ISO 9000 was based. When applying these principles, however, top management should pay more attention to the human aspects (software) rather than just the methodologies (hardware). Creating an environment for good quality culture and developing down-to-earth practices in tune with these principles are as important as applying the principles. Tam et al. (2000) also shared the view that culture related factors are the most important ones affecting construction quality. Writing in the context of the construction industry, the following quality culture must be developed before the industry can reap the benefits resulted from the implementation of quality management systems.

1. In the construction industry, users of the final constructed work are usually considered as customers. For a typical construction project, there may be hundreds of processes involved from conceptual design phase to project completion. Strictly speaking, each process owner is the "customer" of the owner(s) of the preceding process(es) because process owners are internal customers within the quality management system. If each process owner satisfies the owner(s) of the succeeding process(es) (internal customer(s)), the ultimate user (external customer) will be satisfied. Major parties involved in the design and construction of a project include engineers, project managers, main contractor, subcontractors and suppliers. In addition to completing their individual assigned work, developing a culture of satisfying their internal customers is the first step towards total customer satisfaction. This is important particularly for the subcontractors of a construction project. Generally speaking, the main contractor is the customer of each subcontractor

because of the contractual relationship. However, the culture of satisfying their internal customers is usually ignored because there is no contractual relationship between subcontractors. A subcontractor (formwork, for example) does not treat his next subcontractor(s) (concrete and/or steel) as his customer(s). The quality of each subcontractor's work relies heavily on the main contractor's supervision and coordination because there is no incentive for subcontractors to perform better other than just satisfying the minimum requirements of the main contractor. The quality culture of satisfying internal customers should be promoted by the management so that a more operative and harmonious environment will be created among the process owners.

2. Process approach and continual improvement are two main themes of the year 2000 version of ISO 9000. A description of these themes was given in the previous chapters. To create a culture of process approach, the meaning of process must be understood not just at top management level but also at the field level by the crews. After knowing what processes meant, field crews should have a clear picture of their respective processes. Moreover, it is very important for employees to have a basic understanding of other related processes so that the interaction between processes and the impact of one on the others can be apprehended. For example, a formwork worker may not apprehend the reason why a missing bracing can have such an impact on both the cost and schedule of the concrete and/or steel subcontractor(s) who is the formwork subcontractor's customer(s). The task will be more carefully executed if he understands the process of placing formwork and its impact on the process of concreting.

Improvements on processes can be made only if process owners have a thorough understanding of their respective processes. Developing a culture of process approach is therefore the first step towards the continual improvement of processes. Before the implementation of any programme to measure processes and to identify areas for process improvement, the top management must be able to convey the purposes of making process improvement and its benefits to all employees so that there will be incentive to do so. The main purpose of making continual improvement in processes is to increase their effectiveness and efficiency so as to allow the processes to become more competitive in the market. In construction, more communication between the main contractor and the subcontractors is encouraged so that a culture in seeking improvements together in the materials and methods used for a process can be created. Any benefits resulted in such process improvement should be shared so that employees are motivated to continually improve their respective processes. Likewise, such culture should also be developed between the design team and the main contractor.

3. Since quality culture is described as the collective attitudes and beliefs of employees towards quality, it cannot be developed without a focus on teamwork and cooperation among the employees. In construction, adverse relationship often exists among the design team, the main contractor and the subcontractors due to the differences in their interests. As discussed in Chapter 11 on CPCM testing, subcontractors are always to blame for their non-conformities to specifications while main contractors are always to blame for their lack of overall coordination and control. The design teams, on the other hand, are always under the pressure to be more considerate and practical in their design. This culture of blame always exists and the situation may become worse especially when the project is behind schedule or is exceeding the budgeted cost. This is harmful to the spirit of the employees. In general, employees will work with a teamwork spirit and in a cooperative manner only when there are tangible benefits such as bonus sharing for early completion or rewards for excellent performance. Otherwise, it is rather difficult to develop a culture of teamwork and cooperation because of the tight schedule and budget of construction projects.

4. Organizations that promote openness to changes and sharing of information often exhibit good quality culture. Unfortunately, these characteristics are rarely found in the construction industry due to the high competitiveness among the different parties involved in construction projects. With the current information technology, it is not difficult for designers, main contractors and subcontractors to share information in various areas such as construction technology, standard detailing, material and equipment quotations and constructability. This sharing of information among the different parties may be able to reduce some duplicating works and therefore result in a lower total design and construction costs. In the manufacturing industry, this cultural change in information sharing between departments of an organization may be spontaneous because employees can anticipate the result of such change — cost reduction. In construction, however, project participants are often reluctant to share sensitive information because the savings resulted, if any, could end up in their competitors' pockets. Sharing experiences in non-conformances of works is even more difficult as it

will expose the incapacity of the parties involved and may result in higher cost of rework. Moreover, information should be shared not just among the project participants but also the researchers in the industry. Obtaining information from construction companies is very often difficult for researchers with no connections and as a result, many research projects have to be abandoned. In order to make advancement in various areas (such as quality, safety, environmental, tendering and scheduling), the industry must be more open in information sharing and be able to work in collaboration with researchers. Develop a good culture of information sharing between project participants is often difficult because of the conflicting interests existed. However, the concept of project partnering may provide a natural environment to develop such culture.

5. One of the most important elements in management is decision making. In order to minimize faults and ineffectiveness in the decision-making process, it is imperative that decisions made by different levels of management are based on the analysis of data and information. In general, with the help of many well-established methodologies and software available in the market, important decisions in construction (such as tendering, budgeting, scheduling and controlling) are commonly made based on scientifically derived data. However, the culture of collecting and analyzing non-conformance data has yet to be developed in the construction industry. In construction, the collection of non-conformance data is not common and is limited to certain construction processes for which inspections are performed. Such non-conformance data collection should be extended to cover other areas such as design and procurement. On the other hand, the collection of non-conformance data usually creates negative feedback from site workers because the data obtained from such is usually used only for finger-pointing and corrective actions. In order to develop a good quality culture of non-conformance data collection, the purpose must be made known to all staff so that a more positive view on it can be created. In general, the construction industry views the collection of non-conformance data as a fulfillment of the requirement of quality management systems and seldom uses the data so obtained for analysis. Top management staff should bear in mind that the collection of non-conformance data is the first step to continual improvement. Without the non-conformance data collection and analysis, the areas for improvement would be difficult to identify and therefore continual process improvement is difficult to achieve. Instead of just penalizing the staff for their wrongdoings after collecting non-conformance data, they should be educated and reminded that such data is beneficial to the organization as a whole.

12.4 Top Management Commitment

In addition to cultural changes, the commitment from the top management is generally considered by the industry as an indispensable prerequisite for the successful implementation of any quality initiatives. The quality culture discussed in the previous paragraphs cannot be developed without the full support from top management. Besides financial support, personal involvement from top management is essential. Top management must be involved because they are the ones who create the environment to which all other employees will adapt (Cortada and Woods 1995). Low (1998), Serpell (1999) and Lo (2002) also concluded that the lack of involvement from top management was one of the major difficulties encountered by contractors. Top management must have a proper attitude towards quality accompanied by behaviours before employees can be transformed and are motivated to do the same. Moreover, motivational activities such as rewarding and recognizing employees for their contribution to the successful development of quality culture are also essential. In construction, these incentive activities are rare and the culture of rewards and recognition has yet to be developed, particularly at the site level.

Top management must not view the implementation of a quality management system as an end to quality initiative. On the contrary, it is only a first step towards the everlasting process of continual improvement. A successful implementation of a quality management system of an organization cannot always guarantee good quality products. Developing a good quality culture should be a correct and essential step.

As a brief recapitulation, the role of CPCM is illustrated in Figure 12.1. Quality assurance (many existing quality management systems) can only be considered as a stepping-stone for an organization to implement Total Quality Management. As the emphases of the year 2000 edition of ISO 9000 moved closer to the principles of TQM, continual improvement of an organization and its processes are mentioned in many parts of the standard. Specific guidelines for performance improvement are also given in the year 2000 edition of the standard ISO 9004. While much attention is paid to the measurement, analysis and improvement of processes, specific requirement on the cultural aspect is also mentioned in the standard. To achieve continual improvement, ISO 9000:2000 requires the management to develop both the software (culture) and the hardware (structure) for the improvement process. Clause 8.5.4 recognizes the importance of creating a culture as an aid to ensure the future of an organization and the satisfaction of interested parties.

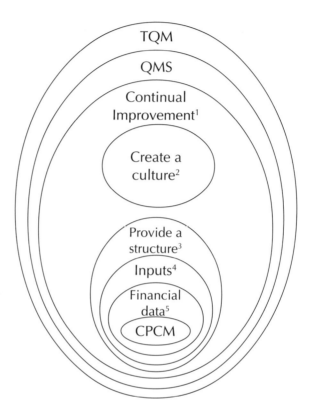

Legend:

ISO 9001:2000 Clause 8.5.1:
1. The organization shall **continually improve** the effectiveness of the QMS through the use of the quality policy, quality objectives, audit results, analysis of data, corrective and preventive actions and management review.

ISO 9004:2000 Clause 8.5.4:
2. ......management should **create a culture** which involves people actively seeking opportunities for improvement of performance in processes, activities and products.
3. **To provide a structure** for improvement activities, top management should define and implement a process for continual improvement......
4. **Inputs** to support the improvement process include......
5. **Financial data** as one of the examples of inputs.

Figure 12.1 The role of CPCM in quality management

In the identification of processes for improvement, financial data is considered as one of the important input from which information is derived. While the previous edition of ISO 9004 offered three models (PAF model, PCM model and Quality-loss model) for collecting and reporting financial data, the year 2000 edition hasn't called out any specific model. After comparing different models proposed by researchers together with feedback from interviews and pilot tests, it can be concluded that CPCM is an effective and practical tool for the construction industry to provide financial data for the purpose of continual improvement of processes. The CPCM approach of quality costs will play an important role in construction related organizations seeking to improve their products in the highly competitive market.

References

Abdul-Rahman, Hamzah (1996). "Some observations on the management of quality among construction professionals in the UK", *Construction Management and Economics*, Vol. 14, pp. 485-495.

Ahmed, S.M. and Aoieong, R. (1998). "Analysis of Quality Management Systems in the Hong Kong Construction Industry", *Proceedings of the 1st South African International Conference on Total Quality Management in Construction*, Cape Town, South Africa, pp.37-49.

Ahmed, Syed M., Aoieong, Raymond T., Tang, S.L. and Zheng, Daisy X.M., (2005). "A Comparison of Quality Management Systems in the Construction Industries of Hong Kong and USA". *International Journal of Quality and Reliability Management*, Vol. 22, No. 2, pp.149-161.

Arditi, David and Gunaydin, H. Murat. (1998). "Factors that affect Process Quality in the Life Cycle of Building Projects". *Journal of Construction Engineering and Management*, ASCE, Vol. 124, No. 3, pp.194-203.

Burati Jr., J.L. and Oswald, T.H. (1993). "Implementing TQM in Engineering and Construction". *Journal of Management in Engineering*, ASCE, Vol. 9, No. 4, pp.456-470.

Carlsson, Matts and Carlsson, Dan (1996). "Experiences of implementing ISO 9000 in Swedish industry", *International Journal of Quality & Reliability Management*, Vol. 13, No. 7, pp.36-47.

Chase, G.W. and Federle, M.O. (1992). "Implementation of TQM in Building Design and Construction". *Journal of Management in Engineering*, ASCE, Vol. 9, No. 4, pp.329-339.

Cortada, James and Wood, John, (1995), *McGraw-Hill Encyclopedia of Quality Terms & Concepts*. McGraw-Hill, Inc., New York, USA, p.102.

Goetsch, David L. and Davis, Stanley B. (2000). *Quality Management Introduction to Total Quality Management for Production, Processing, and Services*. 3rd Ed., Pearson Education Inc., New Jersey, USA, p.165.

Gryna, Frank M. (2001). *Quality Planning and Analysis*., 4th Ed McGraw-Hill, New York, USA, p.216.

Kam, C.W. and Tang, S.L. (1998). "How appropriate is ISO 9000 to construction industry:

Hong Kong experience". *Proceedings of the 2nd International Conference on Construction Project Management*, Nanyang Techonological University, Singapore, February 19-20, 1998, pp.479-483.

Kumaraswamy, Mohan M. and Dissanayaka, Sunil M. (2000). "ISO 9000 and beyond: from a Hong Kong construction perspective". *Construction Management and Economics*, Vol. 18, pp. 783-796.

Laszlo, George P. (1999). "Implementing a quality management program – three Cs of success: commitment, culture, cost". *The TQM Magazine*, MCB University Press, Bradford, U.K, Vol. 11, No. 4, pp.231-237.

Lo, Tommy Y. (2002). "Quality culture: a product of motivation within organization". *Managerial Auditing Journal*, MCB University Press, U.K, Vol. 17, No. 5, pp.272-276.

Low, S.P. (1998). *ISO 9000 and the Construction Industry Practical Lessons*. Chandos Publishing (Oxford) Limited, Oxford, U.K.

Low, S.P. and Omar, Hennie F. (1997). "The effective maintenance of quality management systems in the construction industry". *International Journal of Quality & Reliability Management*, Vol. 14, No. 8, pp.768-790.

Low, S.P. and Yeo, K.C. (1997). "ISO 9000 quality assurance in Singapore's construction industry: an update". *Structural Survey*, MCB University Press, Bradford, West Yorkshire, U. K, Vol. 15, No. 3, pp.113-117.

Mo, John P.T. and Chan, Andy M.S. (1997). "Strategy for the successful implementation of ISO 9000 in small and medium manufacturers". *The TQM Magazine*, MCB University Press, Bradford, U.K, Vol. 9, No. 2, pp.135-145.

Moatazed-Keivani, Ramin, Ghanbari-Parsa, Ali R., and Kagaya, Seiichi (1999). "ISO 9000 standards: perceptions and experiences in the UK construction industry". *Construction Management and Economics*, Vol. 17, No. 1, pp.107-119.

Ngowi, A.B. (2000). "Impact of culture on the application of TQM in the construction industry in Botswana". *International Journal of Quality & Reliability Management*, Vol. 17, No. 4/5, pp. 442-452.

Ritz, George J. (1994). *Total Construction Project Management*. McGraw-Hill, New York, USA, p.10.

Serpell, Alfredo (1999). "Integrating quality systems in construction projects: the Chilean case". *International Journal of Project Management*, Vol. 17, No. 5, pp.317-322.

Tam, C.M., Deng, Z.M., Zeng, S.X. and Ho, C.S. (2000). "Quest for continuous quality improvement for public housing construction in Hong Kong". *Construction Management and Economics*, Vol. 18, pp.437-446.

Tan, Peter K.L. (1997). "An evaluation of TQM and the techniques for successful implementation". *Training for Quality*, MCB University Press, U.K, Vol. 5, No. 4, pp.150-159.

Tang, S.L., Poon, S.W., Ahmed, Syed M., and Wong, Francis K.W. (2003), *Modern Construction Project Management*. 2nd Ed., Hong Kong University Press, Hong Kong, p.104.

Tsiotras, George and Gotzamani, Katerina (1996). "ISO 9000 as an entry key to TQM: the case of Greek industry". *International Journal of Quality & Reliability Management*, Vol. 13, No. 4, pp.64-76.

ABOUT THE AUTHORS

S.L. Tang

S.L. Tang is a faculty member in the Department of Civil and Structural Engineering of the Hong Kong Polytechnic University. He is a Chartered Civil Engineer and obtained has B.Sc. in Civil Engineering from the University of Hong Kong in 1972, M.Sc. in Construction Engineering from the National University of Singapore in 1977, and Ph.D. from the Civil Engineering Department of Loughborough University, U.K. in 1989. Dr. Tang had about seven years of working experience in civil engineering practice in contracting/consulting firms and government departments before he joined the Hong Kong Polytechnic University. He is currently involved in the teaching and research of construction management, and water and environmental management, and has written over one hundred journal/conference papers and books related to his area of expertise.

Syed M. Ahmed

Syed M. Ahmed is a faculty member and Graduate Program Director in the Department of Construction Management (College of Engineering) at the Florida International University in Miami, Florida USA, since December 1999. Prior to this he was lecturing in the Department of Civil & Structural Engineering of the Hong Kong Polytechnic University for over four years. He has a B.Sc. (Hons.) degree in Civil Engineering from the University of Engineering & Technology, Lahore Pakistan and M.S. and Ph.D. degrees in Civil Engineering from the Georgia Institute of Technology, Atlanta, Georgia USA. Dr. Ahmed's research interest is in the areas of project management, construction safety, quality assurance and total quality management in construction, risk analysis and risk management, construction procurement, information technology, and engineering and construction education. He has published over 70 papers in refereed international journals and conferences so far. He is also a reviewer for six international journals in the field of construction engineering and management. Dr. Ahmed has approximately eight years of construction industry experience, working with owners, consultants, and contractors. He is a member of the American Society of Civil Engineers (ASCE), and the UNESCO International Centre for Engineering Education (UICEE) based at Monash University in Melbourne Australia. Dr. Ahmed has successfully supervised three Ph.D. theses and eight M.Sc. dissertations.

Raymond T. Aoieong

Raymond T. Aoieong is a faculty member in the Department of Civil and Environmental Engineering at the University of Macau. He obtained his B. Sc. and M.Sc. in Civil Engineering from the University of Ottawa. Before joining the University of Macau, he worked for consulting firms in Canada as structural engineer for over five years. He received his Ph.D. from the Hong Kong Polytechnic University in 2004 working on the topic "Capturing Quality Costs of Construction Processes using the Construction Process Cost Model". Dr Aoieong is a registered professional engineer in Canada and a member of the American Society of Civil Engineers.

S.W. Poon

S.W. Poon is a faculty member in the Department of Real Estate and Construction at the University of Hong Kong. Before joining the department, he taught at the National University of Singapore and the Hong Kong Polytechnic University. He is a Chartered Structural Engineer and a Corporate Member of the Hong Kong Institution of Engineers. He obtained his M.Sc. in Construction and PhD from Loughborough University, U.K. Dr. Poon was the Chairman (2000/2001) of the Safety Specialist Group of the Hong Kong Institution of Engineers and has been a Senior Member of Professional Committee of Construction Safety, the China Association of Construction Industry, in Beijing since 2000. He has been appointed by the Legal Aid Department and other law firms as an independent safety expert in investigating construction accidents and failures. His research interests include construction and project management, temporary works design and construction, and failures during construction. He has published many papers on construction safety and accidents. He is a co-author of a self-study package of Project Management published by the Hong Kong Polytechnic University and the book entitled *Modern Construction Project Management* published by Hong Kong University Press.